4 ^{es}

DECISIONS
DECISIONS

DECISIONS
DECISIONS
GAME THEORY AND YOU

by ROBERT BELL
and JOHN COPLANS

W. W. NORTON & COMPANY, INC.

NEW YORK

Copyright © 1976 by Robert Bell and John Coplans
First Edition

Library of Congress Cataloging in Publication Data

Bell, Robert I
  Decisions, decisions: game theory and you.
  Bibliography: p.
  1. Decision-making. 2. Game theory. 3. Inter-
personal relations. I. Coplans, John, joint author.
  II Title.
  BF441.B395   158.1   75-23142
  ISBN 0-393-01121-6

Published simultaneously in Canada
by George J. McLeod Limited, Toronto

Printed in the United States of America
1 2 3 4 5 6 7 8 9

CONTENTS

The Lord is subtle, but he isn't simply mean.
Albert Einstein

My God, My God, why hast Thou forsaken me?
Jesus Christ

DECISIONS
DECISIONS

1 THE DEAL

Game theory is about making decisions, especially when one is uncertain of other people's intentions. The focus of this book is on everyday personal problems, particularly private ones, which often provoke irritation, anxiety, and uncertainty. A wide range of such situations, difficult to grasp, sometimes unpleasant, and hard to resolve, can be mastered by the application of game theory.

Game theory has rarely been applied to ordinary and complex private-life problems. Paradoxically, when dealt with in terms of game theory, personal problems which may at first seem baffling (and therefore complex) can be simply and quickly resolved once their underlying structure is understood.

Game theory was invented in the early 1920s by the Hungarian-born mathematician John von Neumann. Anticipating the disastrous course of Nazism, he left Germany, where he was a young professor, and came to the United States. While at the Institute for Advanced Studies in Princeton, he collaborated with another noted refugee from Europe, the economist Oskar Morgenstern, to produce the classic book *The Theory of Games and Economic Behavior,* first published in 1944. The book was intended for economists

1

and social scientists, but it was also of obvious interest to military planners because of its rigorous treatment of the concept of strategy.

This book follows the original framework of game theory, going from simple to more complex problems. It should be read in sequence, chapter by chapter, because each example progressively elaborates the theory. Where possible we avoid the use of unnecessarily complicated technical language; thus some technical terms have been changed, but without altering the basic theory. Those readers familiar with the theory will recognize our usage. However, we hope that the advanced reader will find some of our applications to be of interest.

Though game theory is a very practical method of solving a wide range of human problems, the process of rigorously clarifying and reorganizing one's thoughts may sometimes appear to be inhuman because of the emphasis on logical behavior. However, the purpose of game theory is to avoid foolish errors of one's own, and to anticipate correctly the actions of others—which necessitates the use of logic. But the proper use of game theory does require the inclusion of personal and subjective feelings, including morality. Game theory can only be as moral as the person using it. If after applying the methods described in this book the reader still has moral doubts as to a particular solution, then he or she cannot have correctly assessed all the relevant aspects of the problem, which must then be re-evaluated.

The theory is extremely flexible and allows for the inclusion of eccentric or idiosyncratic desires. If you would rather make love than earn $10 million that's your affair, because everything has a price, even the notion of price has a price. People are willing to pay for the privilege of competing to win even if winning is really losing. This is a game of status, like a potlatch (in which the Kwakiutl tribe of American Indians used to compete among themselves by seeing who could throw the biggest party and destroy the largest amount of his own personal property during the celebration). Money and value are not the same thing; the potlatch quality in our society of competing, for instance, to pick up a restaurant tab is obscured by the cultural tradition of being a generous host. This issue of the difference between value and money is taken up in the chapter "How Heavy Are Your Feelings?" which deals with, among other things, debt as an idiosyncratic fear.

The book helps you define your desires and get them as well by avoiding actions that lead to frustration. The chapter entitled "I Couldn't Do It; I'd Vomit" exemplifies this.

The chapter, "How Do You Feel?" deals with knowing what you want. How can people define their objectives? Money, power, love, or even survival itself may be at stake. On the other hand, the accumulation of money, or the acquisition of power, or love itself may not be your problem, which could be instead to lead the kind of life you want. In short, one's goal may be to obtain maximum freedom and flexibility in one's life style. To do so we shall teach the reader to decide how to rank preferences. Sorting preferences entails knowing what the alternatives are—the obstacles to be faced, the options which are open, which can be successfully taken and those which are traps. All of our examples deal with these problems in different situations.

Sorting out the options can be an intricate job. This may involve talking to many people, and looking around hard for alternatives. The danger is that the search may compound the mental chaos by presenting many more options than can or need to be dealt with. This is common. Often, the more a problem is studied, the more possibilities present themselves. The chapter "Mapping the Conflict" presents a systematic way to organize as many options as required, to avoid a jumbled state of mind. The method presented in this chapter does one more thing: It forces the reader to take account of crucial issues which might otherwise be ignored or overlooked.

The entire book deals with very common and ordinary crises which often tear peoples' lives and emotions. A crisis is a pivotal moment, when the direction of one's future is determined. These are of two types: (1) decisions in which the time element is well known and expanded, where there is time to prepare and think (see "Split Decisions" and "How's the World Treating You?"); and (2) decisions in which there is very little time available, where it is vital to find the right answer quickly (the remaining chapters). A crisis may also be the moment of the showdown between opponents as in "Showdown with Colonel Blotto."

In any personal crisis one must marshal resources. In short, your desired goals, or the threat to your freedom, and how you act is crucial. For a man or woman to just drift along without realizing that time can never be regained and that lost opportunities will be regretted is a classic situation, especially when one wants a changed life style, but fears to make the jump. Dreaming of success may be more acceptable than facing the possibility of trying and failing, as in the chapter entitled "Flashpoint" where a person deliberately creates a crisis to force himself to stop drifting.

Overreaction to a problem, later regretted, is dealt with in "Prisoners of Love." This is an example of "irrational rationality"—where being too rational and not consulting with other people leads to a trap. The absence of trust combined with false signaling produces an unhappy, self-fulfilling prophecy.

Underreaction, or an inability to respond, is the same as endlessly weighing the possibilities—of continuing to research instead of acting. However, some situations call for further study—and by study, we mean trying to find out what the other person(s) is going to do. Alternatively, some situations simply don't call for more information, thus seeking it is a way of procrastinating. Too much information can complicate the problem by creating alternatives that are not needed, or can raise issues which need not be considered. See "The Trouble with Harry" for a discussion of this type of problem.

Not to be forgotten is that some situations inherently lack a clear course of action, and for the reader to be torn between several alternatives is quite usual, and sometimes inevitable. Often there is no "best" decision, yet game theory offers a decisive solution by focusing on the issue of probabilities, or odds, in surprising ways.

This book is basically nonmathematical, so don't be alarmed at the mention of odds, which we use as a guide to action. However, for those readers who want to know the mathematical nuts and bolts, we have included the detailed solutions in parenthetical remarks. These can be ignored without peril. On the other hand, many chapters include important diagrams, which should not be ignored. They are easy to follow if read in conjunction with the text.

Is there some way of assessing the value of the information and deciding what bearing it has on the particular problem one is trying to solve? Yes, depending on whether, without further information, the situation has or has not a clear best course of action. "Showdown with Colonel Blotto" illustrates the importance of secrecy. At times the denial of information is of crucial importance. In this case, neither opponent knew or could predict the other's actions until the showdown.

Another commonly experienced situation is the relentlessly repeated crisis in which one's feelings remain the same, but the pressures of everyday life continually change. Whether or not and when to see a psychiatrist is a typical situation dealt with in "How's the World Treating You?" A very subjective plight is brought under control by the application of game theory.

"To Be or Not to Be a Pawn . . ." deals with countering the cultural compulsion of giving other people the edge by being overly good-mannered. Anyone who has a compulsion has no strategy but to follow it. He has only one strategy, but other people may have more than one.

Regrets are often thought of as irrational and irrelevant. Yet who doesn't have regrets? Sometimes they can be anticipated and included in the decision-making process. An example where regrets are almost the whole ball game is "What Might Have Been."

Up to now we have talked about one's own priorities. Sometimes, however, situations arise which for some reason one is forced to play, and though one has the option of setting one's own values, these are very predictable, as for instance, criminal penalties. Most people want to avoid jail. One still has options, but one is forced to be involved in the situation willy-nilly (see "The Prisoner's Dilemma"). Somehow one has to do as well as one can given the values that are imposed by the law. In "The Prisoner's Dilemma" the system plays on a person's fears to ensure compliance.

Finally, there is the issue of withdrawal as an option. The question is: Under what circumstances can one withdraw?—where withdrawal may be a less complex situation than proceeding with a particular course of action, as in "Colonel Blotto's Retreat." In this example, by withdrawing at the right time in the face of likely but not certain defeat, Blotto converts the likelihood of defeat into the possibility of victory. This is vastly different from withdrawing in the face of certain defeat. Why? Because, if one is certain to lose, withdrawal is an empty gesture.

What about bargaining, coalitions, threats, and other aspects of game theory? The basic idea of game theory is covered in this book. More advanced topics applied to everyday life would be the subject of another book. However, for the reader wanting further information on these topics, we suggest the first two items in the Bibliography, which, by the way, lists items in order of importance rather than readability.

2 HOW DO YOU FEEL? — THE LOWERARCHY

The individual rather than the theoretical person is the focus of this book. It is from this stance that we begin to look at the theory of games. We suggest that game theory is a tool like any other tool, and as a tool it does not function until used. The skill of the person using it is a necessary adjunct of control.

The first axiom is that given a set of alternatives, one can rank them in order of preference, from what one wants most to what one wants least. This is the traditional approach. We propose a point of departure which permits a more flexible and intuitive application. It is our contention that preferences should be stated in *reverse* order, and ranked by beginning with what one desires least of all. In terms of logic this makes no difference, but in terms of human desires and anxieties, it makes a great deal of difference. People know what they do not want. Examples: Survey data for elections invariably reveal that people make their choice by voting against someone, preference being indicated, more often than not, by negative choice. Again, people commonly say, "I don't know anything about art, but I know what I like." This statement is usually provoked by a painting they can't stand. The correct statement is, "I don't know anything about art, but I know what I don't like.'' If a man or a woman leaves his or her parents' house, he or she doesn't say, "I've moved into my own place.'' The common phrase is "I've left home.'' When a husband and wife split, neither says, "I'm living alone now''; each says, "I've left him or her.'' From the earliest age we know what we dislike. Whether healthy or not, a child knows that it hates spinach. Similarly, we believe that in any situation the most effective way to set up a hierarchy of preferences is by a negative process, to begin at the bottom, and state what one does not want. But the word *hierarchy* connotes ranking from top down rather than from bottom up. Since we shall initiate the reverse procedure, our term for ranking will be *lowerarchy*.

The philosopher William James believed that one can never know or discover truth by sitting in detachment in an armchair. Are we advising our readers to sit in detachment in an armchair, evaluate their existence, make key decisions that will model their existence before they live it? We are not saying that at all. We are saying that given the nature of most people's lives as they live them from day to day, decisions have to be constantly made or deferred; anxieties and problems accumulate, sometimes to such an extent they provoke a crisis.

To solve any personal problem, begin the most pragmatic way possible: Rank your order of dislikes. This approach is based on your own gut feelings. With this procedure you immediately begin moving further and further away from what you do not want. (In the movie *Five Easy Pieces,* Jack Nicholson says, "I may not know where I'm going, but I sure as hell know what I'm running away from.'') On this basis, we are dealing with a practical rather than a theoretical approach.

A second axiom is now crucial: Dislikes are transitive, or consistent. If B is worse than A, and C is even more objectionable than B, then C is also worse than A. This may seem deceptively trivial. But without consistency, your preferences are circular. Example, if you dislike eating at home more than eating at restaurants, which you dislike more than eating at your mother's, which you dislike more than eating at home—no matter what you do, you'll be unhappy—or starve. Conversely: If you dislike eating at your home (a) more than eating at restaurants (b) and you dislike eating at restaurants (b) more than eating at your mother's (c), you'll need less bicarbonate of soda (giving mother's cooking the benefit of the doubt) and make mother happy. This also applies to two alternatives. In the following examples the consistency is obviously absent: A man is with Carol but wishes he were with Susan, and when he is with Susan, he wishes he were with Carol. A woman is with Jim but wishes she were with Richard, but when she is with Richard she wishes she were with Jim. When you are working for one firm you wish you were working for another, but when you finally are working for the other firm, you now wish you were working for the first. Endless rotation!

People capable of making decisions do not have a problem with intransitive dislikes; therefore ranking preferences or priorities by the use of a lowerarchy will present no problems. More than that, each time they do something, they will understand the purpose of what they are doing. And having established a lowerarchy of preferences, they will make a decision most suitable to themselves.

Thus we can only make effective decisions and get what we want—or not get what we don't want—if we have a consistent set of values. This is what the use of a lowerarchy is all about. We do not claim this to be the way most people think. Far from it. However, we do claim that this is the way decisive individuals think.

So far, it would appear that all we have said is that effective persons know what they want. Not so. Effective persons know what they do not want. If they occasionally need to iron out any rotations in their thinking, they do it. Knowing where to start is at least half the battle in decision making. One of the objections often raised against this approach is that a person must be single-minded even though human personalities are multifaceted, with a variety of traits. If the traits do not conflict, this objection doesn't hold. Furthermore, multifaceted personalities express their preferences at different times. It is possible to simultaneously like yachting, the opera, movies, and baseball; yet you can't do them all at the same time. (There are, however, problems of conflicting and mixed desires of a different kind which we will deal with in a later chapter.)

3 LET'S GET MARRIED

Game theory is extraordinarily flexible. It takes into account the fact that every action presupposes a context, and that the context is always changing. Here is an example in which whether to get married or not implies a wide variety of confusions.

For the past year Joe has been having an affair with Sue. Joe's parents have been eyeing the situation unhappily, his mother commenting, "Girls today give themselves away too cheaply! Why don't you settle down and get married?" Finally Joe says to Sue, "Maybe we should get married," which throws her into sudden turmoil, since she fears getting trapped.

Here are Sue's options without regard to her preferences: Marry Joe? Marry anyone? Live with Joe without marriage? Continue dating? Break up with Joe? Delay? Any of these options can be considered as a possible response to Joe's sudden proposal. This list gives Sue six possible strategies. Against these, Joe may have responses of his own. For the moment, however, assume that Joe has already chosen his strategy, which is marriage.

Sue's problem: to decide which of the six options she dislikes the least. Of course, each option in itself is complex,

and multifaceted. If Sue accepts the notion of marriage, she
has to decide under what conditions she is willing to marry.
She can marry for one or more reasons, among which are
love, status, emotional security, or children.

Sue's lowerarchy can easily be established by posing
and answering questions in the form of an internal dialogue.

"Are you against getting married?"

"No, but I don't want to marry Joe just for the sake
of his parents."

"So marrying Joe is worst for you?"

"Yes, for now it is."

"Does this solve the problem?"

"No, but it seems to me I'm willing to accept one of
the other options. Maybe I'm willing to live with Joe,
and in any event, why not delay making any decision?"

*"Delay seems quite low on your lowerarchy. It seems
to remove some of your anxiety. But at this moment,
you definitely do not want to marry Joe, just the
thought of it makes you nervous?"*

"Yes."

*"If he says, 'Okay, Sue, how about living together,'
what then?"*

"I don't think I want to."

*"So, we have established that what you least want is
to marry Joe, and what you next least want is to live
with him."*

"In fact, I wouldn't live with him under any circumstances, because that would be the same as marriage."

*"Okay, so they are equivalent in your mind and in
top place. Are they interchangeable?"*

"Yes. Living together and marriage involve many of
the same kinds of problems."

*"Your remaining choices are to continue dating,
break off, or delay."*

"I'd like to keep on seeing Joe rather than breaking off."

"So, now you have to make the final choice between breaking off or delaying."

"I'd rather delay. I'd like to go on seeing him and at the same time delay."

"But, Sue, it can't be at the same time. We are establishing a lowerarchy. Do you want to break off, or do you want to delay?"

"I'd rather continue dating him even if it means dodging the issue."

"To continue dating with the clear understanding of no change is your bottom-line choice in this lowerarchy?"

"What do you mean—bottom line?"

"Okay, the thing you most want. Then your bottom-line choice is to keep things going as they are. We now have the top, next to the top, and the bottom of your lowerarchy."

"All I have are those three choices?"

"No, we are going to get the entire lowerarchy. We still have to answer the question of breaking off or delay. In other words, you prefer breaking off to living with Joe and you prefer breaking off to marrying Joe?"

"Yes."

"So, your second preference from the bottom in this lowerarchy must be delay. Like a rocket countdown, this is your lowerarchy: (4) Marry/Live with Joe (3) Break off (2) Stall, and keep dating (1) Continue dating without commitment."

Sue, having established her lowerarchy, must not go back on it. At this point, to compromise her preferences is to compromise her overall situation, leading to her eventual unhappiness. She now knows that her choices are defined in her order of happiness. If Joe says, "At least live with me," and she shifts her ground, she will distract herself from her real feelings. Of course, over a period of time the lowerarchy may be subject to re-evaluation. At this point Sue mustn't shilly-shally; her lowerarchy is established for the immediate present and the intermediate future, in other words, for purposes of this decision only. Otherwise she violates one of the fundamental assumptions of the theory, the consistency of preferences. If she does, she cannot use game theory.

Note that making a lowerarchy merely reveals how Sue's mind is arranged and is not the same as making a decision. Why? Because a decision has to be expressed in the world by action, and Sue may not be able to get the bottom line on her list.

In this example we have laid out the method to establish the lowerarchy. First, list the widest possible range of alternatives. Second, find the item on the list which causes the greatest anxiety; this becomes the top item. Third, look at the remaining items and find the next that seems to be most objectionable. If you can't make up your mind at once, try pairing two items. Fourth, continue this process until the entire lowerarchy is completed.

Most of this book will be about the method of getting as close to the bottom of your list as possible, or alternatively, staying as far away as possible from the top of your list. Establishing a lowerarchy is merely the first step, since the outcome, what you actually get, depends not only on what you do, but also on the other person's decision as well. Therefore, any final decision must take the other person's decisions into account. Before we get into interdependent situations we shall look at a few more examples of establishing a lowerarchy.

4 HOW HEAVY ARE YOUR FEELINGS?

Although we have established Sue's lowerarchy in the marriage crisis, we have said little about how much she prefers continuing the present situation over delay, or how much she prefers delay over breaking off. All we have at the moment is a lowerarchy. Perhaps all the items, except the last, are bunched at the top; or perhaps all but the top item are bunched at the bottom; or perhaps they are evenly spaced; or perhaps some are bunched and others evenly spaced.

Sue definitely doesn't want to marry Joe, and she definitely wants to continue seeing him—the two extremes. What we want to establish is, To what extent is the present arrangement preferable to breaking off with him? And to what extent is breaking off with Joe preferable to marrying him? In short, exactly where is the breakoff point? So what we want to know is the intensity of her feelings. We won't find this out by simply asking her; if pushed, Sue might say, for example, "Well, I want to break off twice as much as I want to get married." We know quite well that the reply is too arbitrary. Why? Because the answer implies a greater introspective sophistication than can be expected from anybody. Sue has no refined method to externalize in words the delicate intensities of her feelings. We now provide one: Pretend that Sue walks into a Las Vegas casino, and Jimmy the Greek, the famous odds maker, is asked to give odds on Sue's problem with Joe. Jimmy the Greek offers Sue, on the one hand, a choice of a sure thing of breaking off with Joe or, on the other, an even money gamble: a 50 percent chance of marrying Joe against a 50 percent chance of continuing to see him under the present circumstances. (A sure thing of breaking off versus a fifty-fifty gamble on the extremes.)

"Which do you choose, Sue? The sure thing or the gamble?"

"I would rather break off; I don't want that gamble."

So Jimmy the Greek changes the odds. "Okay, those odds weren't good enough for you. Now, it's a sure thing of breaking off with Joe, versus a three to one gamble: a 75 percent chance of continuing the present situation, or a 25 percent chance of marriage. Which do you want, Sue, the sure thing or the gamble?"

"I still don't want the gamble—I'll take the sure thing, breaking off."

Jimmy the Greek tries again: "Okay, I'll make it eighteen to one: a 95 percent chance of continuing the present arrangement versus a 5 percent chance of marriage, against the sure thing of breaking off. Which will it be?"

"No deal, I'll still take the sure thing!"

Jimmy the Greek persists, "Tell you what I'm going to do, a 99 percent chance of continuing the present arrangement, versus a 1 percent chance of marriage, against a sure thing of breaking off. Which do you want, the gamble or the sure thing?"

"That gamble's okay, I'll take it."

Somewhere between the rejected and the accepted gambles is an indifference point (which is where the exact breakoff point lies). Let us assume it occurs with a gamble of 97.5 percent on the present arrangement versus 2.5 percent on marriage. The important point is that the Jimmy-the-Greek method reveals the enormous distance between breaking off and marriage, and the short distance between breaking off and the present arrangement.

Thus for Sue, breaking off, while less desirable than continuing the present arrangement, is nevertheless very close to it. And breaking off is also far preferred over marriage, completely refuting her earlier statement "I want to break off twice as much as I want to get married." Sue knew the direction of her feelings but had no way of expressing them with any accuracy. The true statement is close to forty-five times as much. (See Figure 1.) Of course, if Sue is a compulsive gambler, she will only confuse the issue, and may, to her surprise, end up marrying Joe.

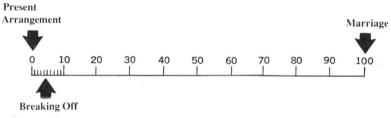

FIGURE 1

Why is the intensity of preferences important? We shall later run across many situations in which a simple lowerarchy of preferences provides insufficient information upon which to make a decision, and intensity, expressed in numbers, will be required. And numbers, by their very nature, refer to intensity. It may never be necessary for the reader to use the Jimmy-the-Greek odds method, but even a rough application of this system will give a good idea of the intensity of preferences and, therefore, what the relative order of magnitudes of any numbers should be. Furthermore, knowing the intensity of your feelings is important for its own sake. To know this gives you a complete picture of yourself in terms of your values. If it seems like a lot of work, remember, a major life decision may be riding on it. And no one can tell you how strong your feelings are except you yourself.

Even though you can compare your intensity of feelings with those of another person, the information is useless. Why? It is misleading and logically impossible: What is worth zero for Sue may also be worth zero for Joe, yet these zeros are not necessarily the same, because with feelings there would seem to be no absolute zero. There may be, but we don't know. The intensity Sue attaches to not marrying Joe may be a much stronger (or weaker) feeling than the intensity he feels about marrying her. And Sue's intensity of feeling toward what is at the bottom of her list, the decision she feels least unhappy about (continuing dating Joe), may be a much stronger or weaker feeling than Joe's bottom line on his lowerarchy.

Thus, since end points are not necessarily the same, they cannot be compared. Later in this book we'll look at the consequences of such a comparison to see if it can be used as a guide to action. Any conclusion based on comparison of feelings is necessarily weaker, because the assumptions are greater. When we do make a comparison, it will only be with the assumption that this is plausible.

While, clearly, we deny that feelings can be categorized into any common denominators, we do claim that we can describe individuals and individual feelings. Again, remember there is no such thing as being "normal." The norm is a phantom. Few can agree on the exact meaning of the words defining it, especially when expressed as an average, which can signify three different and usually unequal numerical quantities—the mean, median, and mode. And the distribution around the mean—what statisticians call the "standard deviation"—is critical, but rarely mentioned. But what is clear is who you are. This does not depend on the meaning of words or the choice of a numerical average. It depends on your personal values.

What about a shotgun or enforced decision—one made under pressure and against our better judgment? Bothered by our conscience, we are more comfortable assuaging it. What we lose out of one pocket, we gain in the other. Your personal compulsion—cultural, moral, religious—may be so powerful that it overrides all other feelings. Don't fight it.

Suppose you hate to be in debt (for whatever reason). Maybe there was a long history of debt in your family, which you cannot forget, and you become compulsive about it.

You buy a car. You can take out a loan and pay the debt every month or you may have an asset you can sell to get the cash for the car. Though this asset may be far more valuable in time than the cost of the car loan, you prefer to give up the possibility of the increased value rather than to be in debt. In short, you prefer peace of mind to money. Contrary to bank ads on TV, money may not be the same as peace of mind to you. In economic terms, this preference is madness, but in terms of your lowerarchy, it is perfectly rational. So, though this example lacks common sense, it's perfectly rational Remember: your rankings are always right because they reflect you.

Every person must be his or her own judge, not executioner. To get through each day you may have to maneuver from one position to another, fundamental anxieties must be taken into account, and integrated into your lowerarchy. Obviously, there are extreme compulsions that cannot be dealt with in this way—alcoholism, chronic gambling, kleptomania, and so forth. But the person who fears indebtedness should not say, "I must overcome this weakness so I can make more money." He or she should say, "No, this is the way I am, I'm going to base my life on recognizing my fears and integrating them into my decision-making process." Why? The reason is: for you to be satisfied with the decision you eventually make, the lowerarchy must be true to you.

5 I COULDN'T DO IT; I'D VOMIT!

In general, one cannot maximize two different functions at the same time. At odds in the following example are the individual desires of a husband and wife. The husband has been offered an executive promotion in another city at a greatly increased salary and responsibility. He decides to take it regardless of whether his wife will go with him or not.

As the husband explains, "Maybe I can later reconcile with my wife, but I know damn well that if I throw away this job, I've lost my career. I may never have another shot at it again. I'm no different from any other guy. If you ask a man to describe himself, he gives his occupation before he says whether he is married or not. He doesn't say, 'I'm a husband,' he says, 'I'm an accountant.' Divorces are frequent, thrown away careers are rare."

Faced with this *fait accompli* the wife now has to decide where she stands. Her list of options: go along happily, go along and be miserable, go along and stake out new terms for marriage, demand that her husband change his mind for their children's sake, separation, divorce.

Here is the wife's dialogue with her alter ego:
"What is it that you really want to avoid?"

"For the family to break up."

"You're sure?—I don't believe you. What do you want out of this? You want your husband to stay? Is that what you really want, for him to stay and be unhappy?"

"No, I want something for myself."

"You feel that this marriage hasn't been rewarding enough?"

"Right. This job offer is just another typical example of marriage denying me my identity."

"Denying you your identity?"

"Absolutely! I might go, if there were some way in which I could have the freedom to stake out my own identity and do the things I want to do."

"Well, what is it that you want to do?"

"It isn't that I know for sure . . . being married you just can't do what you want. I'm constantly tied to my husband and my children."

"Well, what do you not want?"

"I don't want to be pushed around anymore. This job is the last straw. The last thing I want is to be dragged to some strange town as another piece of baggage."

"Even though your husband would be making better money and have a more important career? You may become a wealthy woman in five or ten years. Are you willing to throw this away now?"

"No! Yes! No! Money might provide eventual freedom, so that I could begin to lead my own life."

"But you don't want to encourage him to take the job as things now stand? In other words, you don't want to be the loving, dutiful wife and say, 'Darling, that's wonderful, when do we move?' "

"I couldn't do it, I'd vomit."

"What about the idea of going along, suffering?"

"No, I have one life and I won't do that."

"Essentially, all the options relating to going along, such as going along unwillingly or taking it in your stride . . ."

"I won't go! I need time to think about it, I'd like to delay."

"But your husband doesn't have any time."

"I've just got to have time! I'm not going! He can do what he likes—he can stand on his head. I'm not going!"

"You have gotten ahead of yourself. You have discovered that going along cheerfully, or remorsefully, is undesirable in terms of your own dignity. You have also realized that you can't ask your husband to delay. What about invoking the welfare of the children?"

"I can't use my children as a weapon."

"This is not an option for you?"

"No!"

"But if it is an option, is it at the top of your lowerarchy? It comes above agreeing cheerfully to going along?"

"Yes."

"So the fact is, you don't want to use the children as a weapon. Even if your husband stayed, you would always feel guilty about that?"

"Guilty, schmilty—I have to live with my children."

"Really, worse for you than going along as a dutiful wife is using the children as a weapon. That is what you hate most. All the others you have ranked must be moved down one place. You have three options left in the lowerarchy which we have not yet ranked: letting him go without an official breakup, and deciding later after he is there. Another is divorce. The third is going along and changing the terms of the marriage."

"I don't want to divorce him now."

"If you changed the terms of the marriage, what would you ask for?"

"That's difficult to answer because he gets his own way all the time. I just can't see that there can be any different terms to our marriage if I go along."

"But what kind of terms do you want? Do you want money in your own name?"

"Money is always useful."

"In other words, up to now all you have had are the second set of charge plates. What you want is a large amount of money in the bank in your own name?"

"That's a possibility. Then I won't feel an absolute dependent. What I really want is a stake in the marriage on a fifty-fifty basis. At present there's no way of me getting it."

"Do you know that for sure?"

"Well, I could let him go alone, and make my mind up later about what I want."

"But that would leave hanging the question of divorcing him."

"Why not?"

"If your husband offers you a large bank account and your own checkbook, you still dislike this more than staying exactly where you are?"

"Yes, that's true."

"So, in fact your third preference is resettling the terms of the marriage, not outright divorce. In other words, you would rather divorce than resettle the terms of the marriage?"

"Yes."

"So you are going to let him go alone?"

"Yes!"

"Finally, your bottom-line choice is to let him go alone and leave hanging whether you will go later or not at all."

"I may or may not decide to follow him later. This leaves me free to decide what I really want, to see what it is like without him."

Finally, here then is the wife's lowerarchy:

5. Demand husband refuse job for children's sake.
4. Go without changing terms of marriage.
3. Go, and resettle terms of marriage.
2. Divorce.
1. Husband goes, wife may or may not follow later.

She may cry, but at least she won't vomit.

Comment: Game theory functions with a basic honesty. It reflects only as much integrity as is put into it. Despite the wife's appearance of being wounded and lost, and with no obviously desirable alternative, she makes her first wrenching move to assert her independence by establishing her lowerarchy.

6 MAPPING THE CONFLICT

The conflict map charts the opposing attitudes of two parties, grid-indexed for all possible outcomes. It is like reading a map. If you pick up a road map and look up the name of a street in the index you might, for instance, find "Wall Street, square 1-E." You look across row 1, down column E, and where they intersect you will find Wall Street. The conflict map does the same thing for all combinations of options. The horizontal rows represent one person's options, and the vertical columns represent the other person's. The different outcomes are located at the intersection of the columns and the rows. (Figure 2 shows sample conflict maps.)

Is there time in a crisis to construct a map? In most crucial situations, yes. Setting up a conflict map slows you down long enough, if only for a few minutes, to consider all the options confronting you and the other person(s). It provides a method of listing the other person's available options, as well as your own, and all the different outcomes that result from the various combinations. You choose an option, the other party chooses one, and you and the other party get the outcome that results from the combination. Figure 3 provides a simple example, based on all possible combinations of you

26

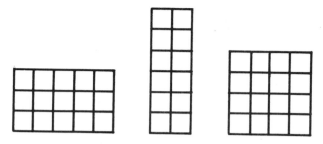

NOTE: Conflict maps are always rectangular.

FIGURE 2A

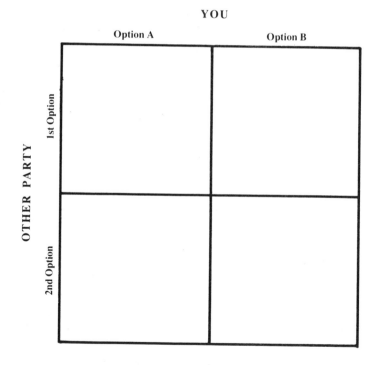

FIGURE 2B

YOU

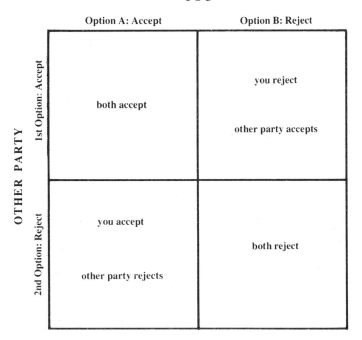

FIGURE 3

and another party accepting or rejecting a proposal (of any kind).

There is no foolproof system of listing all the options open to both parties, but the conflict map is the most reliable system of considering all available information. A conflict map plots a field of expected moves; therefore, you are less likely to be taken by surprise by another person's actions.

What about the versatility of the structure? It allows for any conceivable combination of options by listing all logical possibilities. Extremely flexible, it can accommodate any and all separate options. Moreover, options that blend into one another can be broken into smaller bites. The primary assumption in the use of a conflict map is that both parties make simultaneous choices. If, as often happens, one person makes his or her choice first, or the information changes, the conflict map will require alteration. The important point about both persons making simultaneous moves is that each chooses in ignorance of what the other is going to do. This is the common situation. Thus the question is not who chooses first, but who knows what and when.

Conflict maps will be used in all future examples. In constructing a conflict map it is imperative to use the following procedure: In every situation, start by listing your options and the other person's options in any order. Next, count your options and make a series of columns to fit your options, then count the other person's options and make the appropriate number of rows intersecting the columns. The rows must be sufficient in number to match the other person's options. The results will be a grid such as those shown in Figure 2. This grid will include a space for every logically possible outcome resulting from your options and the other person's. Next, rank only these outcomes according to your lowerarchy. Finally, enter the numerical rankings on the grid, which, when completed, becomes the conflict map.

7 LIZ VERSUS THE SNEAKY PARKER

Liz has found an empty parking space on the street and has pulled her car forward, parallel to the car in front, preparing to back in. Suddenly, a Sneaky Parker shoots into the space. Liz is left momentarily and illegally double-parked. What does she do? She has two options: (1) leave her car double-parked, get out, and complain to the Sneaky Parker; or (2) drive away and look for another parking space. These are her only legal options. However tantalizing, ramming the Sneaky Parker's car, or punching him in the face, is illegal, and, unfortunately, impractical.

The Sneaky Parker has three options: (1) lock his car and walk away fast; (2) stay and argue, "I saw this space first!"; or (3) stay, apologize, and give up the space, saying, "I'm sorry, I thought you were leaving." (The blank conflict map is shown in Figure 4.)

Although the problem of the Sneaky Parker is annoying, it is obviously trivial. However, it provides a good example of how to construct a conflict map of a universally recognizable situation, and to apply some key principles of the theory of games. Of course, there is no time to do this in the actual situation, but since Sneaky Parkers are so com-

29

monplace, there's no harm in rehearsing your own feelings, and comparing them with Liz's.

Liz versus the Sneaky Parker has six possible outcomes. Liz is the column player with two options: (1) drive

FIGURE 4

away, or (2) complain (both options are designated at the top of the map in Figure 4). The Sneaky Parker, the row player, has three options: (1) walk away, (2) argue, (3) apologize and give up the space (his options are designated at the side of the map in Figure 4).

We shall now establish Liz's lowerarchy, in the form of a dialogue between Liz, the rightful parker, and her other oolf:

> *"Okay, I've listed all the results of this situation in Figure 5, a list giving all combinations of your options against those of the Sneaky Parker. The question is what do you least want to do?"*

FIGURE 5

Liz	Sneaky Parker
complains	walks away
complains	stays and argues
drives away	walks away
drives away	stays to argue
drives away	would have given up space
complains	gives up space

NOTE: The list is already in lowerarchical order. See text for full discussion.

"Stay to argue but the bastard walks away—that would be maddening. I would blow my top, and have nobody to take it out on."

"Okay. You complain and he stays and argues with you?"

"That's equally bad!"

"In other words, it's a scene?"

"Yes, and I hate a public scene. I mean, sometimes

people are impossible—you never know who you're going to be arguing with.''

"These two outcomes are of equal value to you, and at the top of your lowerarchy?"

"Definitely. Two things I don't want!"

"These have equal value, are interchangeable, and you don't want either one. If you complain and the Sneaky Parker apologizes and gives up the space?"

"Oh, that would be marvelous, if it could happen. That's at the very bottom of my lowerarchy, the best."

"We have the top two and the bottom items. Now for the others. When you drive away, the Sneaky Parker could have been prepared to argue or walk away?"

"I don't see how I would be affected by what he would have done—I'm driving away."

"So, you're indifferent to these two outcomes, but they are better than complaining when he obviously won't give up the space?"

"Sure!"

"How about when you drive away, catch a glimpse of the Sneaky Parker, and you think that he is reasonable, and quite possibly would have apologized and given up the space?"

"Too late, it's the same to me as the previous two outcomes."

"You are indifferent to all driving away outcomes?"

"Yes, they're all second best."

It is now possible to fill in the complete conflict map with Liz's lowerarchy. Although there are six possible outcomes, there are only three different values Liz attaches to the different outcomes, since she is equally indifferent to some of them. To Liz, the worst thing that can happen is to complain and the Sneaky Parker either walks away or stays and argues. These two outcomes are at the top of her lowerarchy, in the third position. The outcomes associated with Liz driving away were all the same for her, and in second place. The outcome she disliked least was to complain and have the Sneaky Parker apologize and give up the space. Liz's lowerarchy has only three levels:

3. Liz complains, Sneaky Parker walks away or argues.
2. Liz drives away, Sneaky Parker walks away, or stays to argue, or is willing to give up the space.
1. Liz complains, Sneaky Parker apologizes and gives up space.

The complete conflict map, including Liz's lowerarchy values, is shown in Figure 6. Note that the numbers do not indicate intensity of dislike, they are merely used in this example as a convenient way to show the order of her lowerarchy.

Liz makes a wise and clear decision based on this lowerarchy. In grabbing Liz's parking space, the Sneaky Parker showed no concern for anyone but himself. Realizing that the interests of the Sneaky Parker are diametrically opposed to her own, and not wishing to give him a further opening, Liz acts in a way independent of his good or bad intentions. She makes her decision on the basis that no matter what the Sneaky Parker does, he can't cause her further annoyance.

Anticipating that the Sneaky Parker will continue to maximize her discomfort, Liz now acts to minimize it. This is called a minimax strategy. Therefore, Liz should choose the option which has the "smallest maximum" number in it. The largest (worst) value in the column of the conflict map (Figure 6) labeled "complains," is 3. The largest (worst) value in the column labeled "drives away" is 2. Obviously, since 2 is less than 3, 2 is the "minimum maximum"—the minimax (the best of the worst for Liz). Therefore, the option of driving away is Liz's minimax strategy, called in game theory a good strategy.

Liz can reasonably predict that the Sneaky Parker will act defensively and maximize his own situation, to defend himself against any possible move by Liz. (This is called a maximin strategy.) Therefore, the Sneaky Parker will choose the option which has the "largest minimum" number in it. In the row of the conflict map (Figure 6) labeled "gives up space," the smallest number is 1 (best for Liz but worst for the Sneaky Parker). For the other two rows the smallest number is 2 (second best for both Liz and the Sneaky Parker). Therefore, Liz can figure that the Sneaky Parker will either choose to stay and argue or to walk away quickly, both options giving the same maximin value of 2 (best of the worst for the Sneaky Parker). The Sneaky Parker thus would not give up the space. This choice would be the Sneaky Parker's good strategy.

Notice that the minimax value of 2 (the best of the worst for Liz) and the maximin value of 2 (the best of the worst for the Sneaky Parker), are the same. This is an important fact when it occurs on a conflict map. Situations with this quality are called strictly determined. The strategies for each person are clear and unambiguous. As we shall see in a later chapter, not all conflict maps will be strictly determined.

In this conflict situation between herself and the Sneaky Parker, Liz intuitively felt the odds were overwhelmingly against her, so she drove away to look for another space. We have treated the situation through Liz's feelings rather than through any notion of external logic (such as two people being unable to occupy the same space at the same time). To Liz, her own feelings are the real reality.

This example illustrates another important point about game theory. Game theory is not the same as lifemanship, one-upmanship, or notions of this kind; it is a way of realistically dealing with the world in terms of your own feelings.

LIZ

	Complains	Drives Away
Walks Away	3	minimax (best of worst for Liz) 2 maximin (best of worst for Sneaky Parker)
Stays and Argues	3	minimax (best of worst for Liz) 2 maximin (best of worst for Sneaky Parker)
Gives Up Space	1	2

SNEAKY PARKER

LEGEND: 3 = worst for Liz if she complains
2 = worst for Liz if she drives away
So, 2 is the best of the worst, the minimax value for Liz

FIGURE 6

Occasionally, you will get the best of a situation, then the theory may produce results which look like one-upmanship, but as this example illustrates, the results don't have to be "winning." Liz did well on her own terms just by driving away.

We have worked out Liz's lowerarchy in this situation, but we have not worked out yours or anyone else's. Quite possibly, others might have very different lowerarchies, depending on their characters, which could lead to very different decisions from Liz's.

8 THE HIDDEN ADVERSARY—LIZ VERSUS THE JURY

Sometimes a decision must be made whether or not to take action after a legal wrong has been committed against one.

Liz, a married woman, has been raped. The facts of the rape need not detain us. All that matters from our standpoint is that Liz was raped, that she immediately reported it to the police, who rapidly apprehended her assailant. He pleads innocence, claiming that he had sex with her at her invitation. It is a case of the accused rapist's word against Liz's.

When Liz talks to the prosecuting attorney, he informs her of various salient facts about the trial situation: Without supporting evidence, in a rape case it is difficult to get a conviction from a jury. Because of the adversary courtroom situation, the defense attorney will try to turn the case around, into a trial of Liz—a hysterical woman accusing a man of rape; there was no rape—perhaps seduction, perhaps Liz led him on.

The assistant district attorney talks to Liz about the case:

"Liz, you really want the jury to convict this guy, right?"

"Of course, more than anything! He's got to be punished. If he's not jailed, he's going to do it to some other woman."

"Liz, how would you feel if the jury finds the man not guilty? In rape cases without corroborating evidence, most defendants beat the rap. The defense attorney will try to turn the defendant into the victim, and you into a vengeful seducer."

"But, he raped me!"

"I'm not for one minute suggesting that this didn't happen, but you've got to understand that many rape complaints are false. On the other hand, the mere fact that a rapist has been unsuccessfully prosecuted sometimes incites him to more rapes. I have nearly as much interest as you in nailing this guy, but we must have a solid case.

"Liz, his attorney will put you on trial. Every sexual detail of your past may be dragged into open court. We don't know whom they will call to the stand, and it could get pretty ugly. The case could make the local papers. I'm not trying to discourage you from going through with it. On the contrary, but to get a conviction, I have to have a firm and determined witness. Otherwise, there is no chance, it's not even worth bothering with the trial. It comes down to this: How do you feel about pursuing the case and losing, or if we get a hung jury, the whole case repeated in court again?"

"I want him convicted, that's what I most want! If we win, I don't give a damn what comes up in court, I'll be vindicated. But if we lose . . . isn't there any way of holding the trial in private?"

"Absolutely not! The defendant has a right to and will demand an open trial."

"Will I have to describe the rape?"

"Of course. And you even will be asked if you enjoyed it."

"And we could still lose? I'll be mentally raped at the trial, and I could still lose! Well, what other kinds of questions will they ask me?"

"They'll probably ask you every detail about your sex life."

"But I don't have to answer."

"You'll be under oath."

"I can refuse."

"If you show the slightest hesitation, I can assure you the jury will find the man not guilty, because there are no other witnesses—it's your word against his. I warn you, he's entitled by law to the benefit of the doubt. These trials usually last two or three days, especially if the defendant has a good lawyer who will attempt to shake your testimony by every means he can."

"But surely it's just a matter of me giving my evidence."

"We have no idea of his defense other than he is sure to try to turn your testimony against you. Do you realize the implications of a hung jury or an acquittal?"

"What?"

"That you imagined the whole thing, that you are a sick, neurotic woman—that's what will be in the newspapers, maybe even in your husband's mind, that you tried to nail an innocent man."

"I never thought of that. You mean that if I go to court and they find him not guilty, they will think me neurotic or crazy? That's ridiculous."

"That's a very real possibility. What I'm trying to say to you is something very simple. You're justifiably a deeply angered woman. Your reaction is normal. I can't say to you we won't get a conviction—we might, but our chances are slim. If the jury gives the defendant what the law says they are supposed to give him—the benefit of the doubt—they'll acquit him. What I want you to do is go home, and decide if you want to put yourself through this public torture and humiliation even if the jury does not convict. The only reason you should pursue this is if you don't give a damn whether the jury convicts or not, but you simply want to try for justice.''

Liz's decision is made no easier when, after leaving the district attorney's office, she talks to her best friend, who says, ''Everybody knows you've been raped, and the guy has been caught. You can't back down now, you'll be worse than a laughing stock.''

So Liz sits down in complete confusion and works out her lowerarchy and the conflict map (Figure 7):

Liz's lowerarchy:

3. Liz prosecutes/jury acquits
3. Liz prosecutes/hung jury
2. Liz drops complaint/jury acquits
2. Liz drops complaint/hung jury
2. Liz drops complaint/jury convicts
1. Liz prosecutes/jury convicts

Liz's feelings about all the different outcomes are that she doesn't like any, including that of the man being found guilty when she goes through the adversary court situation, because of its ugliness. However, this is value 1, the ''best'' of a bad lot. The worst—value 3—is to pursue an adversary court situation, have her private life dragged out in court, and then watch as the jury acquits the defendant, or is unable to reach a verdict, which could mean a second trial. Intermediate between these—value 2—is to drop the case and forget about what might have been. Liz's good strategy is, therefore, to drop the complaint, which is indicated in the right-hand column.

The minimax value of 2 appears in all three boxes of the right-hand column. This value can only be obtained by choosing the option ''drop complaint,'' which becomes the minimax strategy, the good strategy. The maximin value is also 2, but only is a maximum value in the two top right-hand boxes. This is because in the top two rows—"not guilty," "hung jury"—the smallest number is 2, whereas in the row labeled "guilty," the smallest number is 1. Since 2 is bigger

LIZ

	Prosecute	Drop Complaint
Not Guilty	*3* worst	minimax maximin *2* second best
Hung Jury	*3* worst	minimax maximin *2* second best
Guilty	*1* best	minimax *2* second best

JURY

FIGURE 7

than 1, 2 is the "maximum minimum," the maximin. Thus Liz's own expectation, based on her own lowerarchy, is that the jury will not convict the rapist. In the light of the remarks of the assistant district attorney, her pessimism is realistic.

Liz rationalizes her decision to her best friend: "Do you realize that they are going to drag out in court all the intimate facts of my life? The local papers will report the case. I'll have to describe in court exactly what happened in every detail. They're going to ask me if I enjoyed it. Can you believe that! No matter what I say he still stands every chance of being found not guilty, or there may be a hung jury. What I desperately want is to pursue the complaint and have him found guilty. But if I pursue the complaint, he may be found not guilty, or there may be a hung jury. That thought makes me sick. But if I drop it and then wonder if the jury would have found him guilty . . . I'd better not torture myself over that. So, if I drop the complaint, maybe I can put the whole thing behind me, and forget what the jury might have done, so I've decided not to go through with it."

Of course, someone with a different lowerarchy may have come to a different decision. But the real implication is that Liz never realized until the crunch whom her adversary was. She originally thought her adversary was the rapist, whom she intended to punish by bringing the whole apparatus of the legal system down on his head. When she actually faced the crisis, Liz realized that her real opponent was not the rapist, but a hidden adversary—the law governing rape and the uncertainties of the jury system.

In comparing the conflict map of Liz versus the Sneaky Parker with that of Liz versus the Jury (Figures 6 and 7), we find that they share a similar lowerarchy and conflict map structure. Obviously these two situations are vastly different. Although both situations involve letting an injustice go by, one great and the other trivial, the theory of games shows a much deeper relationship.

Both situations may well be strategically equivalent

for Liz. Liz versus the Sneaky Parker is a less significant version of the decision-making process that she faces in Liz versus the Jury. The magnitude of the stakes change, though the strategic considerations she goes through are based on the same constraints in both cases. The conflict map does not reveal the psychological differences between the two situations. Is this a fault of the theory of games? No. Paradoxically it is a strength. Everyone knows that situations may be equivalent if what is riding on the outcomes is the same. Often overlooked is that situations may be equivalent even if the magnitude of the stakes are very different. In short, situations are strategically equivalent only if the strategic thinking that goes into the decision making is the same.

The problem, however, is that strategic equivalence cannot deal with the overall moral climate. In one case, a person is not a criminal, the Sneaky Parker, and in the other, the person is a criminal, a rapist. This moral climate, however, is reflected in Liz's lowerarchy; she made these situations strategically equivalent. The strategic equivalence is in Liz's state of mind, which established the same lowerarchy for both situations. The theory of games offers no more morality than is put into it; in this sense it operates with a basic honesty. Liz may not have realized she was imposing a strategic equivalence, but she did display a consistent pattern of action and motivation. In the Sneaky Parker situation Liz drove away because she figured she was going to lose and, therefore, a scene would be a waste of time and energy. And in the rape case Liz dropped the charge because she figured she would also lose, and in the process expose herself to public humiliation and private anguish.

We have outlined two situations in which the strategic thinking is identical—one trivial, the other horrendous. But we must add a proviso. Strategic equivalence is a mathematical operation. It is equivalent to converting Fahrenheit into centigrade, or dollars into francs. The numbers in the conflict map must represent intensity of feeling. But the numbers in our conflict maps (so far) merely represent original ranking, not intensity. We have not applied the Jimmy-the-Greek lottery method to our lowerarchies, to determine the exact distances between the outcomes. However, no matter the intensities in either situation, the choice of strategies would have been the same.

9 THE TROUBLE WITH HARRY

The Hidden Adversary was a relatively simple two-choice crisis. In this chapter we deal with a much more complicated crisis, where the decision maker, Kate, has four options from which to choose. The conflict map plays a more crucial role than before in sorting out her decision. In addition, because of further information, Kate changes her lowerarchy in midstream, which results in a change of plan.

Kate is employed as a secretary to the president of a medium-sized company. She discovers that her boss is embezzling substantial amounts of money. Kate likes her boss and works well with him; he has been helpful and kind to her in the past. Not knowing what to do, she discusses her predicament with her friend, Bill:

> "Bill, my boss has embezzled large sums of money from the firm. What if someone discovers the theft?"
> "Would they also find out, Kate, that you knew about it?"
> "I don't think so, but maybe."
> "There's a chance they might?"
> "There's always a chance, Bill, but how could they prove I knew?"
> "How did you find out?"

"I put two and two together."

"Because two and two came over your desk, Kate?"

"Yes."

"Well, Kate, if it comes out in the investigation that this was going over your desk, someone else might put two and two together, too. So you could go down the tubes with Harry."

"What do I do, Bill?"

"What can you do, Kate?"

"If I go and just report it to the district attorney . . ."

"I don't think that is all there would be to it, Kate, you have to give them something to go on, something specific."

"You mean I have to keep copies, I'd have to make copies of things? It's difficult to prove something like this, isn't it? The people who could prove this very easily are the accountants."

"The thing is, Kate, you don't know if the accountants haven't already gone to the authorities. It's possible somebody else in the company may be suspicious. In that case, if it could be shown that you also knew, they could hang a conspiracy rap on you, even though you didn't get anything out of it and just kept your mouth shut. Doesn't it bother you that you know Harry's stealing?"

"Yeah, Bill, a lot. I like the guy, and want to be loyal to him, but he's stolen large sums of money, and someone is bound to find out. I'm just sitting there and sweating. It's very uncomfortable. If I don't say a word, and they discover it, surely they can also discover that I knew, and I'd be in a lot of hot water for not reporting it. Or I could go to him and tell him I know, and say I'll cover if he makes restitution."

"Well, if he doesn't, Kate? You can just quit."

"Yeah, I could get another job easily and then tip off the district attorney."

"Look, suppose you get another job and also tip off the D.A. and they don't have a clue, you will have informed on Harry. Doesn't that bother you?"

"At least I'd be covered."

"You could be sending a man who has been kind to you to prison. But if it turns out they are investigating this case anyway, then you're not really turning Harry in."

"Bill, I don't know what to do. The whole thing is just one incredible mess. I wish I'd never discovered it. Now it means I have to do something, and I don't know what!"

Kate thinks to herself about the trouble with Harry:

"What's the worst for me in this situation? For the authorities to already know about it and find out that I knew and I didn't tell. Is this the worst thing? Am I sure? No, wait a minute, the worst would be to actively cover up for Harry, and for them to find out. Then I'd really be in trouble. I'd be an accessory."

"If that's the worst, what would be the best thing? To cover for Harry while he makes restitution and the authorities don't find out?"

"I suppose it would be terrific. It would solve the problem of my conscience. But if there is an investigation . . . Yeah, that's the catch. The same option which gives me the best outcome also gives me the worst. It's at both the top and the bottom of my lowerarchy, depending on what the D.A. does."

"Let's see. What about staying on the job and just keeping my mouth shut?"

"I can't do it. I couldn't face Harry. Sooner or later I'm sure I would subconsciously show that I know."

"Would I feel any different staying and saying noth-

ing depending on whether the D.A. investigates or not?"

"No difference; I'm going to be very uncomfortable with Harry. It's second from the worst on my lowerarchy."

"How about going to the D.A.?"

"Yeah, but that would mean ratting on Harry, who has been very friendly and good to me—I find that extremely distasteful."

"But isn't it better than staying and keeping my mouth shut?"

"I'm not going to stay and keep my mouth shut. Sure, it's better. There must be another possibility . . ."

"If I tip off the D.A. and he would not have investigated without me telling them, would I feel terrible about it?"

"Yes, I'd feel guilty for years."

"Suppose they are already investigating, I'd be saving myself then? Isn't that better?"

"One degree better, at least I'd cover myself. Now, suppose I leave and get another job. Then I could abandon the whole mess."

"In other words, I'll feel better leaving rather than going to the cops? Do I care whether the cops investigate or not so long as I leave?"

"Yes, because if I leave and they don't investigate, I've at least run away from the mess. If they do investigate, I might be dragged in as a witness for the prosecution. But it's not my fault; I didn't trigger the whole thing. If the cops investigate, and I have to go on the stand, then obviously I'm under oath, and I'll have to tell the truth. That can wait until the day comes. Right now, I want out."

Kate's lowerarchy:

7. Kate covers up/D.A. investigates
6. Kate stays and keeps mouth shut/D.A. doesn't investigate
6. Kate stays and keeps mouth shut/D.A. investigates
5. Kate tips off D.A./D.A. doesn't investigate
4. Kate tips off D.A./D.A. investigates
3. Kate quits job/D.A. investigates
2. Kate quits job/D.A. doesn't investigate
1. Kate covers up for Harry/D.A. doesn't investigate

While Kate was going through her dialogue and establishing her lowerarchy, she filled in her conflict map, shown in Figure 8. Her good strategy, her minimax, is to quit. This decision gives her the minimum trouble. She makes it by figuring that the police will find out about Harry's theft, and she will be called as a witness.

What we have shown in Figure 8 is that Kate has a sense of loyalty to Harry, providing he repays. This is reflected in the number 1, in the upper left-hand corner. However, because 7 is in the same column, she would rather not take the risk. So it is the second factor which prevents her from covering up for Harry. She uses her minimax to determine that she cannot do what would lead to her most preferred outcome. The lowerarchical rankings by themselves are not sufficient to resolve the crisis for Kate. She needs the conflict map as well. This was also true in "Liz versus the Jury." You can't always get what you want. She deduces that her most prudent course is to quit her job without admitting to Harry that she knows, and without giving him the chance of restitution with her help.

The following day Kate goes to work and tells Harry she is leaving. Harry asks why. Kate is evasive and does not give a satisfactory explanation, so he presses her for an answer; but she continues to be evasive, which makes Harry suspicious that she knows about his shenanigans. He asks her to stay and break in the next secretary. She will not. He asks her to at least give him time to find somebody. "No, I want to leave immediately, I just want a letter of reference from you and then I want to go. I don't even care about my salary check, I just want to leave. In fact, I don't even want a letter of reference." Kate suddenly realizes that if Harry is a crook, and may well be found out, what use would his letter of reference be? Harry realizes that Kate must know, and offers her more money to stay and better working conditions. She still refuses. Harry now knows that Kate knows. He says, "Kate, is there some other factor?" She replies affirmatively, but refuses to discuss the matter.

Kate now realizes that Harry knows that she knows. Sensing this, Harry says, "Come on, Kate, we've been good friends. I've taken good care of you. I'm president of this company; whatever I do is my responsibility, it has nothing to do with you. The best thing for you is to stay here and continue to look after me. I'll continue to make it worth your while."

Suddenly Kate is aware that Harry has been nice to her because he has been stealing from the company; he has been setting her up to use her later, in case she found out. She thinks, "He has been using me!" At this point, her lowerarchy changes: She decides to tell the authorities, even though she intends to leave the company. She has had confirmation from Harry, who has virtually confessed, and offered her a bribe. Saying "Well, I'll have to think about it," Kate goes to lunch. Over lunch she goes through her lowerarchy again, just to make sure. She no longer has the option of staying and keeping her mouth shut. Her options are: (1) collusion with Harry, (2) go to the authorities and quit, (3) quit without going to the authorities. She still does not know if someone else in the company has gone to the authorities, who may already be investigating.

KATE

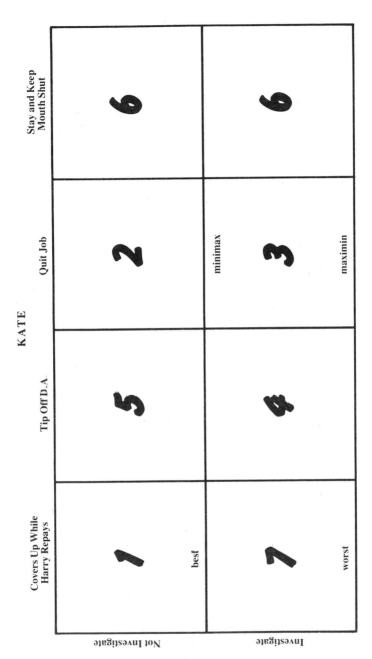

	Covers Up While Harry Repays	Tip Off D.A.	Quit Job	Stay and Keep Mouth Shut
Not Investigate	1 *best*	5	2	6
Investigate	7 *worst*	4	3 *maximin* / *minimax*	6

D.A.

FIGURE 8

This is Kate's internal dialogue after talking with Harry:

"I have only three options now—collusion, tipping off the D.A. or quitting the job."

"Well, I was going to quit the job. Now things have changed. I now see that Harry was kind to me because he was scared that I might discover what's he's been up to. He was just using me. Now that he knows that I know he has been robbing the company blind, he offers to cut me in."

"What about the bribe he offered me?"

"That is just more of Harry's fraud. Collusion with Harry is the worst, whether or not the authorities investigate."

"So these two outcomes are ranked the same?"

"Right. What has changed is that yesterday I thought I owed the guy every kind of loyalty because he had been decent to me. I was just going to run for it. But now, I realize he has been setting me up. He even allowed for the fact that I might discover his antics."

"But if I quit the job, and it came out during the investigation that I knew, I could be in serious trouble. Now I do know about Harry for sure, he has virtually confessed."

"True. Harry might even drag me in if the authorities learn about him. There's no question that if I quit the job, I might still be in a lot of trouble if the D.A. investigates. The best thing for me is to get another job and tip off the D.A."

Kate's lowerarchy:

4. Kate colludes with Harry/D.A. investigates
4. Kate colludes with Harry/D.A. doesn't investigate
3. Kate quits job/D.A. investigates
2. Kate tips off D.A./D.A. investigates

53

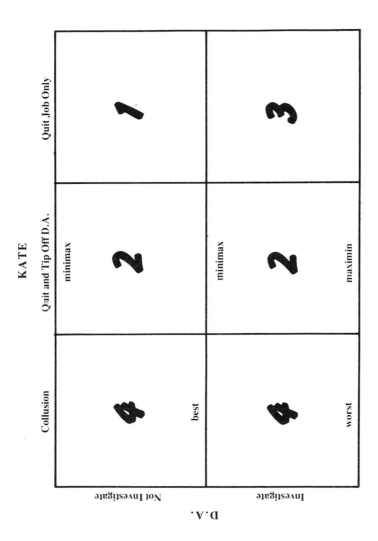

KATE

FIGURE 9

2.   Kate tips off D.A./D.A. doesn't investigate
1.   Kate quits job/D.A. doesn't investigate

The revised conflict map and lowerarchy is shown in Figure 9. Now, Kate's good strategy, her minimax strategy, is to tip off the D.A. She rejects outright the option of collusion, because tipping off the D.A. is clearly a lesser evil. Each value for quitting the job and tipping off the D.A. is lower than the corresponding value of collusion. (This is called a minorizing strategy.) Second, the option of just quitting the job is quite tempting for her, but she decides that the police will investigate. So she must reject this notion as well. This leaves her no choice but to go to the police.

An interesting point is that Kate need not waste her time trying to find out whether the police are investigating or not. She can act now; further information is unnecessary. Game theory states that if a conflict is strictly determined—each player having a pure strategy—further information plays no role.

10 V.D.—LIGHT AND EASY

May has been having an affair with Ken for quite a while now, but not exclusively. One morning Ken phones May and tells her he has V.D., advising her to be checked out that day at the clinic; they do not discuss who gave whom the disease, but arrange to meet for dinner that night.

May has occasionally slept with other men on a casual basis, and realizes that it is possible she has given V.D. to Ken. She doesn't know whether Ken has been sleeping with other women, suspects he has, but isn't bothered. Their arrangement has been implicitly an open, mutual one, but recently Ken has shown an interest in a tighter relationship. They never talk about sleeping with other people. May, in fact, has been deliberately avoiding the subject. She had lived with another man for several years and the relationship became stifling and repressive. Now, although she enjoys the emotional contact with Ken, she feels the need to maintain her distance—at least for the time being.

Thus May is anxious to preserve things as they stand. Since this relationship has been based on not talking about casual affairs, her problem is how to handle the present situation. May's alternatives are to admit or deny that she has been sleeping with other men. She figures that Ken has the same options. If May values openness above all, she has no trouble making a decision. The conflict map in Figure 10 is an example of a lowerarchy based on honesty. In this case, May simply admits, and feels no uneasiness. Her lowerarchy:

2.   Both don't admit
2.   May doesn't admit/Ken admits
1.   Both admit
1.   May admits/Ken doesn't admit

But May does feel uneasy, as revealed by her internal dialogue:

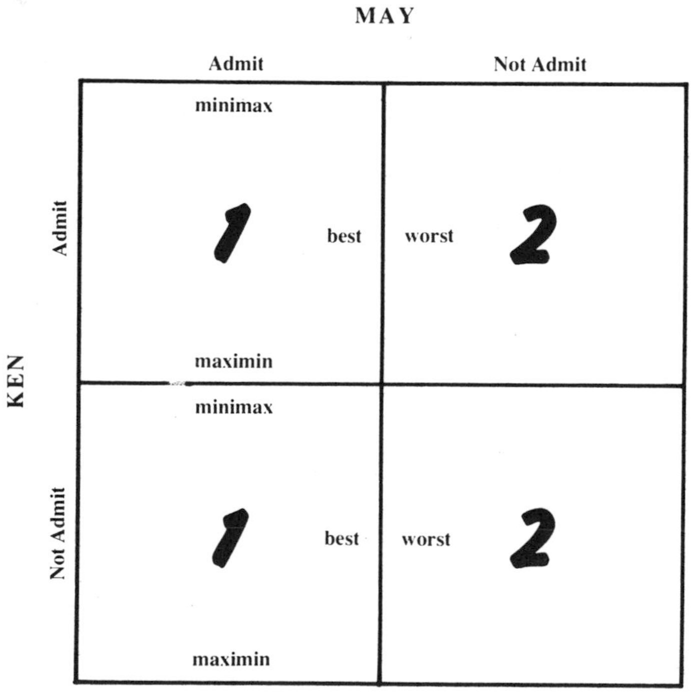

MAY

FIGURE 10

"I don't know if Ken has been sleeping with someone else. If he hasn't been, it will be obvious that I have been—I'm going to appear callous in his eyes. As it stands, we have a first-rate relationship; he doesn't impinge on my freedom, I don't impinge on his. We meet each other whenever we want, go away together on vacations, and have a good time. Now V.D. has complicated everything. It seems to me, the best thing for both of us is to avoid talking about it."

"Why is this the best?"

"If we start to talk about it, and Ken hasn't slept with someone else, he is going to blame me."

"So by neither of you talking about it, you are just not opening that can of worms?"

"Yes, the thing is to forget it, don't have anything to do with it. Be realistic. Okay—either one of us can or will or might get V.D. You go to the doctor and it's cured. Very easy—it's easier to cure than a cold. If I had a cold, and I gave Ken my cold, we would just take it in—no recriminations. Why should we suddenly have trouble over this? The main problem is: I don't think Ken is going to treat it as a cold. Moral overtones. Having a cold is one thing, V.D. is another. It means we've been playing around with other people. I just don't see anything wrong with that, but I don't want a whole lot of moral guilt dumped on me."

"So you prefer that you both don't talk about it. He goes off to the V.D. clinic and gets himself fixed up?"

"No question. It's as easy as can be. There shouldn't be any issue over it."

"What if you could discuss it openly, if Ken were not to throw any recriminations at you, but to admit that he also has been sleeping with other women?"

"Providing he is cool and he doesn't want to inhibit our present style or load it further . . ."

"Load it further, what do you mean?"

"Try and create a completely monogamous situation."

"You still want to be free?"

"Yes—that's been the implicit basis of our relationship as far as I'm concerned. No inhibitions on our freedom of action, what one can do and what the other can't—light and easy. But now Ken might not feel the same way."

"Do you think that if you talk, it won't be light anymore?"

"It would be like living with a guy again! Not to talk about this leaves us free and open."

"Then absolutely bottom line on your lowerarchy is for neither of you to talk about whom you have been sleeping with?"

"Sure—get cured and forget it."

"What about if you were both to talk about it openly. Do you think that would be okay?"

"No, because if we talk about it, Ken might say, 'Well, if we carry on the relationship, then I don't think we should sleep around.' He might try to lay down rules, formalize things in some way."

"You don't know for sure that he will do this."

"More than likely he will."

"What if he says, 'Look, you've been playing around, I've been playing around—neither of us is to blame'?"

"Oh well, if he admits that he has been sleeping around, then I can say, 'Me too, and it's fine—you have your freedom, and I have mine.' "

"Is that of equal value to neither of you admitting?"

"No, if he admits, it would be because he wants to get closer to me—hell, he may not have been sleeping around. He may even say to himself, 'Well, to get closer to May and cement the relationship, I'll admit I've been sleeping around. Maybe I'll even provoke

her jealousy, maybe then she'll agree not to sleep
around.' ''

*"For both of you to admit that you have been sleep-
ing around raises more questions than you want to
answer?"*

"Exactly. Where do you go from there? Closer!"

*"What about the situation where you admit you have
been sleeping around, but he doesn't?"*

"Oh, that would be ridiculous. If I admit and he denies, he dumps a whole load of guilt on me. I'm the
one who's wrong—I've given him V.D., he's pure
and I'm tainted!"

*"Have you ever put any obstacles in his path to sleep-
ing around?"*

"None whatsoever."

*"You think he would be laying a guilt trip on you if
he didn't admit to it and you did? Is that the worst for
you?"*

"Absolutely. The problem is how one avoids all these
guilt things. If I gave Ken V.D., I'm very regretful—
but that's the chance one has to take. It's more than
likely that if I did, he will say, 'I don't want you
sleeping around.' And that will be the end of our
present relationship. Either I'll have to formalize it, a
monogamous relationship leading to living together or
marriage, or we break up. Either way, I don't want
it."

*"So, you admitting and he not admitting is the worst
thing for you?"*

"Yes."

*"What about the opposite—he admits and you
don't?"*

"Again he puts me into a guilt situation. He is now
being generous, and I'm being untruthful, because I
have been sleeping around. He says, 'I have,' but I'm
not admitting anything!"

"Are you saying that it doesn't matter if you admit

and he doesn't, or if you don't admit and he does?''
"No, I'm saying that if he admits and I don't, he claims moral superiority, and he can imply, 'You're a person of a lesser moral fiber.' ''
"In terms of your lowerarchy, where does that outcome fit? Is that of equal value to you admitting and he not admitting?''
"No. I'd rather he admitted and I didn't."
"Why?''
"I don't want to admit because, if I admit, then he can put his foot down and say, 'If you sleep around anymore, I will have nothing more to do with you.' If he admits, he pulls 'honest Ken' on me. But that's better than me being 'tainted.' ''
"In other words, either one of you admitting when the other doesn't is worse than both admitting?''
"Yes. I'm not interested in being Othello in drag. I simply don't want to know if he sleeps with anyone else—all that jealousy and possession crap!"
May's lowerarchy:

4. May admits/Ken does not admit
3. May does not admit/Ken admits
2. Both admit
1. Neither admits

May's lowerarchy has been entered on the conflict map, Figure 11; she now has to make a decision based on it.
She continues her internal dialogue:
"Okay, you've worked out your lowerarchy. What are you going to say to Ken tonight?''
"When Ken phoned, he was agitated. I couldn't tell whether his agitation was about his getting V.D. itself, from me getting V.D. and giving it to him, or from his getting it from someone else and giving it to me. He may think that I want to throw our friendship

MAY

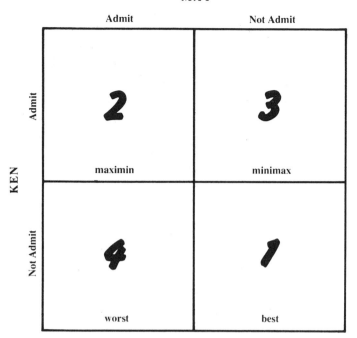

FIGURE 11

away. Also, I can't tell whether he is agitated because both of us having V.D. is repellent. So I can't figure out what his lowerarchy is. I know what mine is, I've just worked it out. Since I've already included my feelings toward Ken in my own lowerarchy, and I do think he wants to cement the relationship, the safest thing for me is just figure that Ken's lowerarchy is the opposite of mine. This may or may not be true but that way I'll be prepared. So whatever is a low number for me would be a high number for him. The only way I can get the bottom line item on my lower-archy—not to discuss—is by not admitting anything. I don't admit, and hopefully he doesn't admit'' (Figure 12).

"But what if he does admit?"

"If he admits, then I don't get number 1 on my lower-archy, I get number 3 (shown in Figure 13). I get the upper right-hand corner of the conflict map, not the lower right-hand corner. Now it's obvious if I think he is going to admit that he has been sleeping with other women, I would be better off admitting as well, because instead of getting 3, I'll get 2, closer to the bottom of my lowerarchy'' (Figure 14).

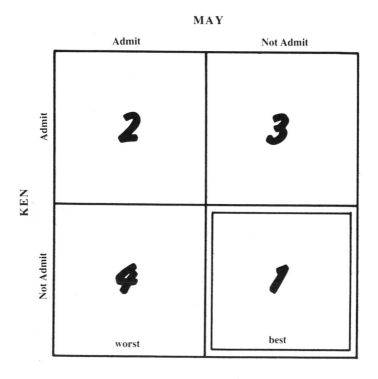

FIGURE 12

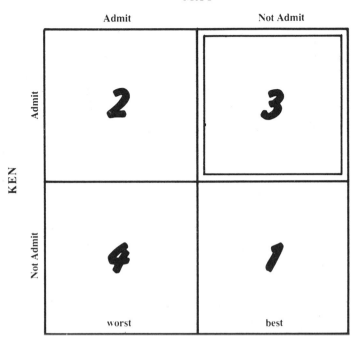

FIGURE 13

"What if he figures that you are going to admit? He would change his choice from admitting to not admitting, putting himself in a better state, but you in a worse one. You wouldn't get item 2 (in the upper left-hand corner), you would get item 4, the worst possible outcome (in the lower left-hand corner.) He gets a more preferable situation for himself—he becomes the one who doesn't admit when you do admit" (Figure 15).

"That's the worst. Having switched from not admitting to admitting, I end up in a trap. Yet if I think he's not going to admit, because he thinks I'm going

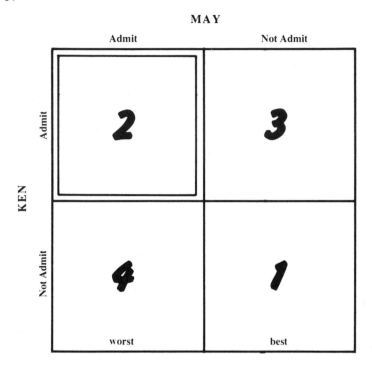

FIGURE 14

to admit, then I should switch my choice back to not admitting, and I get the best outcome on my low-erarchy again'' (Figure 12).

"But if he figures you for this, he might admit, and you're back to item 3 again" (Figure 13).

''Then it doesn't matter what decision I make. Every time I make a decision, it brings up a counterdecision. This is ridiculous! And every time I make the counter-decision, it returns me to a position I have already been in. When I get in a position I was already in, I move to a previous position. I established a consistent lowerarchy, I know what I want, but now I don't

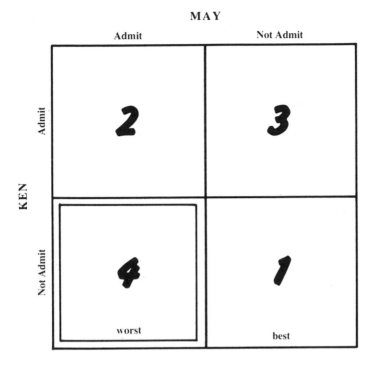

FIGURE 15

know what to do. Now I'm torn between alternatives. The best thing for me to do would be just to wait and see what Ken says. Then, if he admits, I can admit. If he doesn't, I don't. Some relationship! But it fulfills my needs."

"He is obviously going to wait to see what you say."

"It's going to be a long, awkward evening. If he takes the initiative, I can take advantage. If I take the initiative, he can take advantage. If he talks first, I can improve my position. If he admits, I admit. If he decides not to admit, I don't admit. If I know what he is going to do, I know what to do. If I don't know what he is going to do, I can't figure out what I should do.''

May's lowerarchy doesn't lead to a clear decision. Unlike every other conflict map we have looked at, hers is not "strictly determined." The minimax, 3, is not equal to the maximin, 2. May does not have a clear choice. She does not have a pure strategy which is good against anything Ken does. The more she reflects on it, whichever way she chooses becomes unsatisfactory. Any choice she might make based on what Ken might do is good grounds for Ken making a different choice. Therefore, she will choose differently; therefore, he will choose differently; therefore, she will choose differently; therefore . . . Completely circular, there appears to be no way out.

"He thinks that I think that he thinks that . . ." reasoning can't be used because it never stops. But obviously the decision is to say nothing. No—May fears Ken will think her not only a liar but devious as well.

There is an answer. Flip a coin. This is called a mixed strategy. What does tossing a coin do? It fixes a course of action for May. In this morass of indecision it gives her a decision. Since she can't decide which decision to take, the best thing is a commitment to a process of making a decision. And since one decision is as good as another, a random process is best. It guarantees one thing: She stops second guessing Ken. She doesn't even know herself what she will do. Sitting across the dinner table from Ken, she takes a coin, flips it, and then either admits or doesn't according to the way the coin falls.

May could also use a different randomizing device, one which Ken would never suspect. She crinkles a dollar bill. At the moment when she decides to either admit or keep quiet forever, she looks at the serial number of the bill. If the right-hand digit is an odd number, she admits, if it is an even number, she does not. (If it's zero, she looks at the next number.) Ken would never know that May is making her decision randomly.

This randomization procedure supplies May with an anchor, a way of making a final decision, which she can't get from her own reasoning. She announces her decision at the exact instant she makes it. This gives no time to change her mind. She acts decisively. In addition, if Ken knows May well enough to read correctly from her face what she intends to do, he won't be able to, because May doesn't know until the last possible second. She is not only keeping the decision secret from herself—so she can't go back on it—she is also keeping it secret from Ken. This is the principal reason for using a randomization strategy (some game theoreticians think that keeping one's decision secret from the other person, who thus cannot use the information to her/his advantage, is the only reason).

The problem for May is that at the dinner table she may admit first and declare her hand. To avoid this and to take the lead, May could say to Ken, "Look, don't say anything; this whole thing is very embarrassing. The temptation for both of us to be righteous is too great. There's only one way we can resolve it. Write on your napkin 'yes' if you have been sleeping with someone else, and 'no' if you haven't. Don't let me see what you've written. I'll do the same. Then we'll just uncover the two napkins at the same time." May can still randomize using the serial number on the dollar bill. If, in fact, Ken has not been sleeping with anyone else and if they both write "no," either he believes the tin-cup theory of V.D. or he knows the truth.

(The odds of fifty-fifty are the best possible odds May could use based on her lowerarchy, the numbers—4,3,2,1— expressing her intensity of feelings as a guide to action. In the sense of what mathematicians call expected value, May can expect the same amount regardless of whether Ken admits or not. The computation is as follows: Suppose Ken admits, and May uses the odds of fifty-fifty. She gets $1/2(2) + 1/2(3) = 2 \ 1/2$. Now suppose Ken does not admit; May gets $1/2(4) + 1/2(1) = 2 \ 1/2$. Thus May can expect to get the same amount regardless of whether Ken admits or not. If May is a gambling woman, this is the way to assess the odds. The expected value of the gamble is 2 1/2.)

11 SPLIT DECISION— DIVIDING THE DIVISIVE HEIRLOOM

Liz's mother, Sarah, is married to Al, Liz's stepfather. Sarah inherits an antique of some value, and she and Al have the antique appraised at a reputable dealer who makes a substantial offer for it.

Sarah and Al live in the city in an apartment. Sarah, occupation housewife, would like to sell the antique and use the money as a large downpayment on a house in the suburbs, which she intends to leave to Liz. Otherwise she can simply leave Liz the heirloom itself, it being her property. Al does not want to move to the suburbs; he wants to put the money obtained from the sale into his business. Also, he doesn't want to commit himself to traveling to the suburbs and to making continuous payments on the house. He prefers to live in town near his business.

Sarah's lowerarchy and Al's are exactly opposite, and endlessly circular as they argue over what to do. Sarah begins:

"I don't want to sell it unless we put it into a house. I want a house, and would love to live in the suburbs because I'm tired of the city. Besides, I don't want the money put into your business, I want Liz to have access to it. If something happens to you, then there

69

are problems of winding up an estate. Or if something happens to me, I want to be sure Liz will get the money straight away.''

''But, Sarah, Liz is a grown woman. She can take care of herself. She has a good job. We're getting on. I'm fifteen years away from collecting Social Security. We've got to think of our own future. If I put this money into the business, it would mean a substantial improvement. We'll make more money— there may be more money to leave to Liz. But we will at least be taken care of. We've got to think of our own future, Sarah. We've got to worry about what will happen to us in ten or fifteen years.''

''But if we have a house, and if we pay the mortgage payments for ten or fifteen years, we'll own it.''

''Oh, sure, we can own a $50,000 house, and live off $200 a month Social Security. If I put this money into the business now, we could have a substantial income when we retire. We could enjoy ourselves, take it easy, go to Europe, travel around the world, go fishing . . .''

''That's fine for you, Al! I don't want to go to Europe, I don't want to go fishing. I've got fifteen years of my life to live, in a decent house, out in the country with fresh air. I want a garden. I'm tired of the city. I want fresh air and sunshine. I want to be in the suburbs, and I want a house. Now's my chance. If we sell and the cash goes into your business, it could be lost. We can't lose the money if we put it into a house. It's not possible—it's an investment!''

''But what if something happens to me, Sarah, while I'm making the payments—you could lose the house . . .''

''If something happens to you, I can sell the house and I'll have the downpayment back and the payments in between. But it'll be mine, and if something happens to you, if the business goes sour in some way, Liz won't lose the money. This way, I can look out for her, don't you see? If I put the money in your business and something happens to me, I can't be sure and neither can you—the money for Liz will be tied up in the business.''

''Sarah, we don't know anybody out in the suburbs— all our friends live in town. It's just boring in the suburbs. And most of the people are going to be younger, with little kids.''

''Al, you don't take it easy enough. You smoke too much. You're always down at the office. If you lived in the suburbs, you'd play golf, swim, get to know new people and get some relaxation, fresh air. It would be healthy.''

''It would drive me crazy; I'd be the healthiest lunatic in Pleasant Valley.''

''You're just saying that because you want the money for your business. I'm not going to give it to you.''

''But, Sarah, this is a chance for me, maybe my last chance to be somebody. In five years we could be rich.''

''I don't want to be rich, not if it takes a gamble with our one big stake. I know you, Al—you're full of daydreams. You always say we could be rich, you've said it time and time again, but the business has never improved.''

''I've never had this kind of capital before. Sure we could plunk it down on a house, but if we could put it into the business, in five years we could buy an even better house.''

''Al, if that thing is worth money now, it'll be worth maybe more money in ten years' time. It's safe.''

''I think this is a time when we should take a chance. Instead of opting for a safe, secure thing, we should make a play for the big time.''

"If anything happens to you, I'm alone. I need to have control of something that belongs to me. I was left this thing, and I want to be able to put it into something I want. It's mine, Al."

"Look, I've never treated the money I've made as if it was mine and not yours. I've always given you everything you've wanted—the charge plates at the stores . . ."

"Al, this is something special."

"In other words, what's yours is something special, and what's mine is a husband's duty."

"You know that's not what I mean."

"Sarah, you're being unreasonable. I've consulted you on every major decision I've ever made. We've talked over everything. Now we have a chance to really do something worthwhile—not just buy a house and be another anonymous couple in the suburbs—we could really be somebodies."

"Al, I am quite content to be what I am. I don't want to be Mrs. Somebody. If you want to make me Mrs. Somebody, go out and do it yourself, but I'm content."

"You know the problem I've been up against. In my business, you simply can't expand without capital, and this would finally give it to us. I think you should sell so I can put the money in the business. That's the best thing for us."

"Al, I won't sell unless we choose the house, the papers are drawn up, and we put a small deposit down. You sign the papers and I'll sell."

"Sarah, don't be difficult. You've got a good offer on this antique. Who knows, we might not get another offer like it. I don't know anything about these things, but the dealer might change his mind. We are arguing about nothing. Look, until you've got the cash, you can't get the house. Get the cash and then we can talk about it.''

''I won't get the cash until we agree about the house. I know you, Al, you're going to try to get me to change my mind. I'm not going to change my mind; I won't sell until you've signed the papers.''

''I won't sign the papers until there's cash. I'm not going to sign papers for the house, and then we can't raise the cash. I have to lay down a thousand dollars just to sign the papers, and I could lose that if we can't raise the rest. It's not businesslike. Cash in hand is one thing. Look, I'm in business, Sarah. People change their minds all the time. What'll happen is this: When you go to him to sell, he'll say, 'I don't want it now.' And he's waiting for you to say, 'Well, what will you give me for it.' We would be committed to the house, and then we haven't got the cash. We'll have to run around looking for a buyer. Anything can happen. Get the cash, and we'll talk about it.''

''I know exactly what you're thinking. Once the cash is in hand, you can wear me down and put it right into your business.''

''It's the prudent way of dealing with it. Don't commit yourself to anything until you've got the money in hand. You're talking abstractly—you want this house; you don't really have the money. You think you have, but you don't. Get the money and we'll talk. I'm not opposed to the house.''

''You're not opposed to the house? Well, maybe I'll sell. We can phone up that antique dealer, confirm he'll buy, and look for a house tomorrow.''

''Okay, we'll sell and then we can decide what to do.''

''Wait a second—you said you agreed to the house.''

"I didn't say I agreed to it, I said it wasn't such a bad idea. Get the money and we'll talk."

"Well, I'm not going to sell then. You sign the papers on the house and I'll sell. What makes me suspicious is that the only way you can extend your business is with the cash. If you make arrangements to extend your business, and then I decide not to sell, you'll be in trouble. I know what you're thinking, Al. To you, not selling is like losing the money, and buying the house is also like losing the money!"

"Sarah, look, don't sell it. I don't want to move in any event, so don't sell it. But let me get a loan on it for my business."

"If you get a loan on it, and we put it up as collateral, and something goes wrong with your business and the loan is called in, the bank will take it. Nothing doing. Al, I'll sell if you will sign for the house."

"Sarah, I told you I'm not going to commit myself to the house unless we have the cash first."

Al's lowerarchy:

3. Al agrees to house/Sarah sells
3. Al insists on money for business/Sarah refuses
2. Al agrees to house/Sarah won't sell
1. Al insists on money for business/Sarah sells

Sarah's lowerarchy

3. Sarah sells/Al insists on money for business
2. Sarah refuses to sell/Al agrees to house
1. Sarah refuses to sell/Al insists on money for business
1. Sarah sells/Al agrees to house

As shown in Figure 16, the situation between Al and Sarah is endlessly circular, not simply because their interests are totally opposed but because the best decision requires endless adjustment in the light of what each one will do, which provokes further adjustment.* This is because the minimax value is not equal to the maximin value. The antique is such a bone of contention that both Al and Sarah want a resolution. Al needs a way of resolving the circularity of the objective conflict between the two. The conflict is objective because the lowerarchy of one is simply opposite to that of the other. What has happened is that something of value has entered their lives, by chance, and it has produced instability.

Sarah persists in looking for a house, so finally Al says in desperation, "Sarah, what you want is exactly the opposite of what I want. We've got to compromise. What compromise? We've looked at several houses. There was one which you were very enthusiastic about, and if we sold the antique, we could make a substantial downpayment on it. But there was another house we looked at that was smaller, which we both agreed was quite a nice house with a good yard. I will agree to go with you this afternoon to the real estate agent and sign the papers on that house. It won't require as big a downpayment as the first one. In fact, the smaller house only requires about two-thirds of the money we'll get for the antique. I'll sign the paper on this house and pay out the initial deposit of $1,000 before you even sell the antique if you'll give me the other third of the proceeds from the sale for my business: two-thirds for the house, one-third for my business. I think it's a fair deal, Sarah. You agree to sell, I

---

* Worst for Al is for Sarah to sell the antique and he agrees to put all the cash down on a house. Equally bad for him is to make arrangements to put the money into his business, and Sarah decides not to sell. Better than either of these outcomes is, when Sarah won't sell the antique, for him to talk up the house, agree to the house. But best of all is for Sarah to sell and Al puts the money into his business. However, this is worst for Sarah. Almost as bad for her is not to sell when Al agrees to the house. Best of all, however, is to sell when Al agrees to the house. Of equal value is for Sarah not to sell when Al wants to put the money into his business.

AL'S LOWERARCHY

agree to split the money, giving the lion's share to the house.''

Sarah reflects over the proposal and accepts.

What set up the circularity in this situation was that originally Al and Sarah had a stable relationship based on Al as the earner and decision maker. They hardly ever argued. The stability of their relationship was of long standing. The injection of a new element, the heirloom under the control of Sarah, introduced an instability into the relationship. (The same kind of instability occurred in the relationship between Liz and Jim when a new element was introduced—V.D.)

In this example, the solution works because Sarah has two distinct options—sell the antique or not sell it. Al also has two options—put the money in his business or sign the papers for a new house. Al's options, however, are infinitely

SARAH'S LOWERARCHY

AL	Not Sell	Sell
House	minimax **2**	maximin **1** best
Business	maximin **1** best	**3** worst

NOTE: Since they are complete opposites of each other, the problem could have been resolved by using only one of the conflict maps. Notice that in neither conflict map is the minimax value equal to the maximin value; they do not occur in the same square. It is important to remember that whenever this happens there is no clear choice for either person. The resolution of the problem lies in an unusual use of odds, discussed in the text.

FIGURE 16

divisible; he can put part of the money in his business and part of it into buying a house. The division he chooses—two-thirds in the house and one-third in his business—is extremely fair.

Suppose Al had been a gambling man and chosen his final decision using a randomizing device. The best odds, yielding the safest gamble, would have been a two-thirds chance of putting all the money in the house and a one-third chance of putting all the money in the business. These odds would have given Al the same expected value regardless of the choice made by Sarah: Using Al's lowerarchy, we get $1/3(3) + 2/3(2) = 7/3$, if Sarah had chosen not to sell. If she had chosen to sell, Al could expect $1/3(1) + 2/3(3) = 7/3$. With these odds, Al could expect to get the same amount no matter what Sarah did. In terms of a safe gamble, this is the best Al could do. (The same odds would be best if the computation were made on Sarah's lowerarchy.) The situation, however, did not allow for a gamble, but the ratio which would have been used in the gamble—two to one on the house—provided a reasonable division of the money. It turned expected value into actual value. (We are assuming that for purposes of a guide to action the numbers in the lowerarchies reflect the intensity of feelings of both Al and Sarah.) Notice that Al, to be prudent, is forced to put the heavier odds on his less desired option, the house.

Incidentally, if Al had computed his best odds using Sarah's lowerarchy, nothing would have changed. This is because, as is clear in their argument, her lowerarchy is exactly the opposite of his. Thus Al's minimax computed on his own lowerarchy is identical to his maximin computed on Sarah's lowerarchy. Put another way, his lowerarchy on his conflict map is also his hierarchy on Sarah's conflict map, and vice versa for Sarah.

12 TO BE OR NOT TO BE A PAWN . . .

After completing his graduate studies, Fred moved to the city. To his chagrin, he discovers that the academic world is not the gold mine he believed it to be—there is a shortage of college teaching jobs. Halfway through the summer Fred still doesn't have a teaching job for the fall. He's worried, despite two leads for half-time jobs consisting of two "firm" verbal promises: "If the job is available, you'll get it." Both department chairmen warn Fred that at the last minute they may not have a budget for these courses.

Unfortunately, the teaching times for the two jobs conflict, otherwise Fred would be happy to take both. Fred's problem: He has verbally bound himself to accept both jobs. And if both jobs come through, because of the conflict of times, he will have to refuse one, and offend a chairman by breaking his word. Fred feels, in all fairness, that he should immediately withdraw from one job, selecting the one he prefers. In addition, he is haunted by the thought that if he withdraws from one job and the other falls through, he will be stuck without any money. Fred, about to phone and cancel one job, has a queasy feeling about the situation and therefore consults his friend Jim, who is also a teacher.

"Jim, I've gotten myself into a stupid situation.

When I left graduate school, I didn't ask my chairman
to line me up with a job. Instead, I went off to Europe
for six months, confident I'd get a job when I re-
turned. How can I teach European history without
ever having been there? When I came back, I discovered my chairman was away on sabbatical, and I
couldn't find a full-time job because of the tight situation. I ran around looking for part-time jobs. I got two
leads that were reasonably definite, one in the city
college system, the other in the state college system,
but the schedules conflict. I tried to rearrange the
schedules but couldn't. Because of the current funding crisis, both chairmen were unable to make a firm
offer in writing until school started and they were sure
funding hadn't been canceled. But they insisted on a
firm commitment from me, which I gave, otherwise
they would have withdrawn the offers. It doesn't
seem fair, but I guess the chairmen are trying to cover
themselves from extra last-minute work. At any rate,
here I am, a month away from the beginning of the
semester, at the beginning of my teaching career, I
can't afford a bad record, so I've decided that the
right thing for me to do is select the job I want—the
city college—and cancel the state college. Somewhere
in my mind there's a lurking suspicion that this is a
matter of survival. I don't know if I'm just being
timid and foolish.''

''The thing is, Fred, it looks like what you're most
afraid of is telling the chairman at the state college
that you're not going to take his job, and finding out
at the last minute that the job at the city college isn't
going to come through. True?''

''Right!''

''That's the worst thing, as you see it?''

''It scares the hell out of me.''

''On the other hand, what you feel best about is phoning the chairman at the state college, canceling the job, and having the job at the city college come through?''

''Absolutely!''

''What about if you don't phone, keep your mouth shut, wait for the term to begin, and if at the last minute it turns out that the job at the city college isn't available, you'll still have a job?''

''Well, it's obviously better than having no job—I could eat. But suppose the city college job comes through—I mean that's the problem. I would have both jobs, and I can't take both, that's too opportunistic for me to stomach.''

''The system imposes on you a demand to act in a gentlemanly way. It saddles you with responsibility but does not counterbalance the responsibility with the privilege of a job. The state college chairman is merely inconvenienced if you let him down at the last minute, but the chairman still has his job, regardless of what you do. The system is opportunistic, not you, because by imposing a moral obligation on you without guaranteeing you anything in return, it relegates you to the status of a pawn.''

''I suppose that's true—the real issue is one of closing doors—they'll never offer me a job again.''

''So it comes down to this, Fred: Would it be worse keeping your mouth shut and the city college job is available, than keeping your mouth shut and the city college job is not available?''

''Obviously, yes.''

''Okay, Fred, we have a consistent lowerarchy for you'' (shown in Figure 17).

Fred's lowerarchy:

4. Fred cancels state college job/city college job not available

FRED

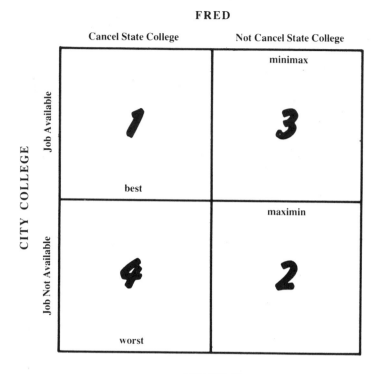

FIGURE 17

3. Fred doesn't cancel state college job/city college job available
2. Fred doesn't cancel state college job/city college job not available
1. Fred cancels state college job/city college job available

"Look, I understand what you're getting at, but I don't understand this map because it only deals with the city, it doesn't deal with what the state college might do—cancel out or give me a job."

"The state college is only a pawn, a hostage. You want the city job. You're tempted to hang on to the state job, just in case. In fact, your fate will be determined by the availability of the city job. The city is the other player in this game, the hidden adversary. In effect, only the city makes a choice as to whether it will make a job available to you or not. It's not the chairman's fault. Obviously, he wants to hire you, but he can't if there's no budget. Surprisingly, your real conflict is with the college you most want to work for."

"Well, what's my course of action?"

"Okay, Fred, look at the conflict map (Figure 17): You most want to cancel out the state college if the city college job is available—there's a number 1 in that box on the conflict map. But you're afraid the city college job might not be available, then you'll get 4 instead of 1, the worst instead of the best. And if that's the case, you'd rather switch your strategy from canceling to waiting, in which case you get 2 on your lowerarchy, and you have the state college job offer. So you think 'what if the city college job is available?' then you'll get 3 on your lowerarchy, close to the worst. So then you think you should switch back to canceling out at the state college. The circularity on the conflict map reflects your own uneasiness."

"Wait a moment—I might not get either job."

"Then it wouldn't make any difference whether you told the state college or not, would it?"

"No, it wouldn't."

"That's why I've left if off the conflict map."

The resolution of this problem is not straightforward because the conflict between Fred's short- and long-term survival is a genuine quandary in his mind. As shown on the conflict map, the maximin is not equal to the minimax, so there is no stable outcome. This is the kind of circular low-erarchy we have run into in previous examples—the V.D. and the heirloom problems. As in these earlier examples, there is a best set of odds, gambling odds. If Fred is a gambling man, he should take four straws, make one shorter than the other three, mix them up, and if he draws the short straw, pick up the phone and withdraw from the state college job. If he draws a long straw, he should wait until the term begins before informing the state college. By using odds Fred takes a slight gamble over his future at the sacrifice of certainty in the present.

(The odds we suggest are the safest because, in terms of expected value, Fred gets the same amount regardless of what the city college does. If the city job is available, Fred gets $1/4(1) + 3/4(3) = 2\,1/2$. If the city job falls through, Fred gets $1/4(4) + 3/4(2) = 2\,1/2$. Fred is satisfied, the lowerarchical numbers express sufficient intensity to be a guide to action.)

But Fred may not be a gambling man. Then the safest course is simply to use his minimax pure strategy, which is to keep his mouth shut. Based on his feelings as expressed in his lowerarchy Fred would regret canceling—"what a fool I am; how could I have been so stupid."

13 HOW'S THE WORLD TREATING YOU?

From time to time Liz has felt the need to see a psychiatrist. Whenever she made an appointment, the therapist invariably could not see her immediately, so she would forget it. One day her need became so unrelenting that she finally decided she had to see a psychiatrist. This time, she did wait and she did see the psychiatrist. The treatment lasted for about a year and although a fairly unhappy experience, she felt it was of immense benefit.

It worked so well Liz decided that she no longer needed therapy and could now relax and enjoy its benefits—which she did, for a while.

During the next two years, Liz went through a series of ups and downs. At times she would decide that she had to go back and see a psychiatrist, but whenever she phoned, she went through the same delay before she could get an appointment. During these periods she would begin to take a much more optimistic view of her situation and decide that the money was better spent bringing her some personal pleasures. But no sooner had she decided against treatment, and canceled the appointment, than she would go through another state of depression, and again ponder whether to go to the

psychiatrist or not. Her life seemed like a pogo-stick ride—
all ups and downs.

Aside from this, Liz could never determine whether
she really needed therapy, especially since her psychiatrist
had never given her a clue about her mental condition.
Whenever she asked, the issue was thrown back at her.

Liz talked over her perplexity with Jim:

"Look, I can't afford a therapist. We can't afford it.
Every time I go see a therapist it's a burden on you,
we can't go away, go out to dinner, and I can't even
dress well—I feel like a poor relative; I have to ask
my parents for money, which is a big hassle and you
don't like it. It's just one big pain in the ass."

"Agreed, Liz, but if you need to go to a therapist,
go."

"By the time I get an appointment, things pass by,
you know, and I don't seem to need it anymore."

"You mean, by the time you see the therapist you
feel better?"

"Right. Everything's all right, I feel the world's treating me better. Then afterward, I get depressed—
things go badly with you and my parents. It's not so
much your fault—it's mainly my parents. And once I
get into problems with my parents, then I get into
problems with you; then I get into problems with my
job. Everything builds up and I get desperate about
the whole situation. So I make an appointment with
the therapist, who keeps me waiting a few weeks
before I can start. By the time I go, everything's all
right again—I've kind of made up with my parents,
with you, with my job, and I'm back where I
started."

"Look, Liz, it's obvious—you're most unhappy not
going to a therapist when things are going wrong for
you. But by the time you see the . . ."

"You don't understand—it's not just a matter of

seeing the therapist—I don't get to the therapist. I know what a boring and painful process it is. Then by the time I'm supposed to see him, I feel more or less fine, so I don't need him. You know, when the world's okay, I simply don't need a therapist.''

"Then the worst for you is when the world is not going your way and you're not going to the therapist?''

"Yes, it's very depressing. You've got to understand, Jim, I have to go to the therapist to get myself up. That's even more depressing.''

"You're not committed to going to a therapist?''

"What's the point when I'll feel better later, anyway?''

"So, you prefer not going when you feel the world is treating you okay, Liz?''

"Obviously. Why should I waste time and money and go through all that pain.''

"So, this is bottom line on your lowerarchy, the best?''

"That's right.''

"And, Liz, we've got the worst thing on your lowerarchy—not going to a therapist when the world is treating you badly.''

"Absolutely true.''

"What about going to a therapist when the world is treating you badly?''

"I need the therapist, and I'm seeing him? Yes, this is second best. At least at this point I'm sensible enough to realize that I can make best use of him, and I do. I explain my problems because I'm right there— with them and him—and I go further into myself. I seem to be able to explain much more clearly what my underlying problems are, and reveal more rapidly to myself, let alone the therapist, what my situation is.''

"Okay, Liz, so that's second best for you—going to the therapist when the world appears to be treating you badly. That leaves one other alternative, going to a therapist when the world is treating you well."

"Yes, but going to the therapist when the world is okay just makes me unhappy. I'm perfectly happy until I get there and begin to delve into myself. I destroy my own confidence. Jim, you've never been to a therapist—you don't know how painful an experience it is, how you get involved with parts of you that become agonizing. I can be in a terrific mood—getting on with you, my parents, my job; I go to the therapist and suddenly something comes up I haven't perceived before that's a mortifying blow. Maybe it's my own stupidity. Maybe I've not been able to interpret a relationship I was perfectly happy about. I thought, 'Well, that was fine.' Suddenly, the reverse happens. I begin to doubt everything."

"Is this as painful for you, Liz, as when you don't go to the therapist and you feel the world is treating you badly?"

"No, it's better. The worst is not going when things are going wrong. But the whole thing is a quandary in another sense: One corner of my mind says maybe I'm making some kind of progress in being able to handle the world, and because I'm better able to handle the world, I won't get so depressed. The problem is that you can go to a psychiatrist for years and years and years before you really get into a stable situation in which you can handle the world. I keep changing. You keep changing. The world keeps changing. And it seems to me an unlimited process of probing into oneself and preparing oneself. It could take five years, ten years, any amount of time and money."

"I see. In other words, Liz, going to a therapist when the world appears okay is better than not going to a therapist when the world in your own eyes appears to be treating you badly. There is the chance of progress.''

"True. But there's another problem you don't understand. Every therapist is a human being. Some people think they are like scientists, that they can simply cure you. They aren't. They get depressed. I've noticed my therapist has bad times. He gets depressed with me. He doesn't know that I know that he is getting depressed, that he has problems. A lot of times, I say to myself, 'Maybe this therapist isn't the right person for me; maybe I'm wasting my money. Maybe I ought to find another one.' You know how you find a therapist? Through friends. Then you're committed to him. The first therapist I went to was a woman. The second one was a man, and there were remarkable differences. Had I only gone to one, I would never have had the kind of perceptions I have now about therapists. The more you go to one, the more sophisticated you are about them. I've been through it all, and sometimes I suddenly say to myself, 'Progress, what progress?' ''

"This question only comes up when you're seeing a therapist and you feel the world is treating you all right?''

"Yes! Also one grows older and is not so naïve anymore; you know a lot about the process. You know how complex it is. And you know that you're a pretty normal person, and that lots of people go through the kinds of problems you're going through.''

"Yeah, I understand. But all of this doesn't change your lowerarchy. It reinforces the lowerarchy.''

"What do you mean?''

"All the different outcomes remain the same: Not going to a therapist when you feel the world is treating you badly is still the worst; not going to a therapist when you feel the world is treating you well is still best; going to a therapist when the world is treating you badly is still second best—then you need him; and going to the therapist when the world treats you well is third best. This is when you worry if you have the right therapist, or whether you need a therapist at all.''

In other words, your lowerarchy is as follows:

4. Not go/world bad
3. Go/world good
2. Go/world bad
1. Not go/world good.''

"How does this help me to know when I go to a therapist? The lowerarchy is how I feel, and it describes my feelings accurately (Figure 18) but it doesn't solve this problem. How does it help me make a decision? I decide one thing one moment, and another when things change. I have no anchor. My conflict map has no stable outcome.''

"But, Liz, there must be a way to solve this problem. You are the one who decides whether the world is treating you well or badly. It's your perception. The world is just going on doing whatever it's doing, and nothing really objective has changed—you have a job, you're still with me, your parents are still there. At times you feel the world is treating you badly, and at other times you feel it's treating you well. But you control the decision whether to go to a psychiatrist or not, and whether to stick with it or not. It's no good flipping a coin because you're not trying to keep your decision secret from the world, you're not trying to outsmart the world. That would be crazy. Furthermore, this is a long-term repeated decision. It's as if you have to make the decision every time you head

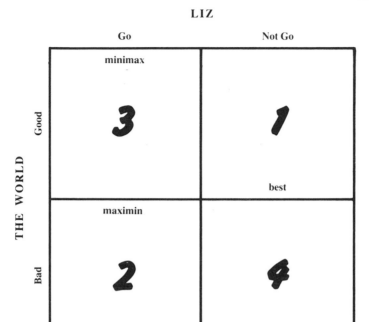

FIGURE 18

for the psychiatrist's office, or every time you stay
home. The problem, I think, is that you feel stuck
with the decision forever. When you're going to the
therapist, you feel that it's going to be forever, with
no clear ending in sight. When you're not going, you
feel that you are completely cut off and totally alone.
Again there is a feeling of finality, particularly since
it's such a big job to start seeing a therapist again.
And to see one, you have to punish yourself with the
expense. So you feel saddled with the choice of
seeing or not seeing a therapist. Another thing, it's no
good one week deciding to go to a therapist, seeing

him a few times, feeling better, then quitting. The problem is that your memory is too short. You forget about the months of agony alone without the therapist. Also, when you're seeing the therapist, and you're feeling bad, you forget about the months before when you were feeling good. You have to deal with this problem in terms of a long memory.

"Liz, you're going to have to look at this in a different way. You go to see a dentist twice a year—every six months—whether you feel you need him or not. I think there is an analogous principle involved."

"How? I don't see it."

"We can set up a schedule for you to see the therapist for a period of months, stop seeing him for a period of months, and then start seeing him again for a period of months. You'll have to clear this with him, of course, and see if he'll go along with it. You could set up a definite schedule. So there would be a period when you see him and a period when you don't see him. Maybe you'll see him for a longer period, and sometimes for a shorter period, but we can work out a definite schedule. It would be the best way of handling the situation."

"I see, Jim. It would be better to plan ahead, with the idea that I'm going to go through all these phases and changes of mind, and that I would be more secure if I went in periodically, independent of whether I'm up or down. And I could have a steadfast inner development this way."

"That's right."

Jim suggests a schedule, taking into account Liz's entire memory of decisions concerning the psychiatrist. Both are aware that Liz's problem essentially involves endless repetition of the problem in Figure 18. Game theory offers a way. We can imagine that this problem is infinitely repeated, which it is for Liz, since there is no clear terminal point. She is continually confronted with making the same decision over and over again, but she has a fixed conflict map. The resolution does not involve the strategic issues discussed in previous examples. Secrecy is not a consideration. Trying to outfox the other person is not an element, since there is no other person than Liz. (The conflict map has no stable outcome because the minimax does not equal the maximin.)

There is an element in this new approach which is strikingly appropriate to Liz's problem—memory. To find the schedule the problem must be fictitiously repeated many times. This is what Liz does (Figure 19a): Suppose on the first go around, the state of the world is bad. Referring to her conflict map, Liz gets 2 by going to a psychiatrist, and 4 by not going.

She starts by dividing a sheet of paper into five columns, the first headed "How the World's Treating Me," the second

FIGURE 19A

How the world's treating me	My total if I go to a psychiatrist	My total if I do *not* go to a psychiatrist	I choose	Play sequence
bad	2	4	not to go	a
bad	4	8	to go	b
bad	6	12	to go	c
bad	8	16	to go	d
good	11	17	to go	e
good	14	18	to go	f

NOTE: The first numbers in the second and third columns (2 and 4) are taken from the bottom row of the Figure 18 conflict map (when the world is bad). Similarly, subsequent numbers are the sum total of these numbers and the respective numbers from either the top or bottom row of the conflict map (depending on whether the state of the world is good or bad).

"My Total If I Go to a Psychiatrist," the third "My Total If I Do Not Go to a Psychiatrist," the fourth "I Choose," the fifth "Play Sequence." She enters the first results: "Bad" in the first column, 2 in the second, 4 in the third, either "go" or "not go" (it doesn't matter) in the fourth, and *a* in the fifth.

On the second go around, Liz merely has to refer to the columns. Recalling from her lowerarchy that 2 is a preferred result to 4, she now chooses, like a smart girl, to go to a psychiatrist. She enters "to go" as the second item under "I Choose" (the fourth column). Suppose on this second go around the world again treats her badly. She enters "bad" in the first column, and now adds the conflict map results of 2 and 4 to the previous entries of the second and third columns, giving totals of 4 and 8 respectively. In the fifth column she writes *b,* meaning the second go around. Now she has filled in two full rows on the table.

On *c,* the third go around, Liz continues to go to the psychiatrist because 4 is a better result than 8. Thus "go" is entered in the fourth column, and since the world is still bad, Liz's new running totals are 6 and 12. On play sequence *d,* the fourth go around, the world is again bad, Liz continues to go to the psychiatrist and her running totals are 8 and 16. On play sequence *e,* the fifth go around, the world starts behaving itself and "chooses" to be good. This is entered in the first column. Referring again to her conflict map, she adds the numbers 3 and 1 to the sum of the last two numbers, giving new totals of 11 and 17 to be entered in the second and third columns, and the letter *e* in the fifth column. Thus on each play Liz should add her new possible results to the total of her old results. And she should always choose on the next play that strategy which has the smallest total associated with it. So on Liz's sixth go around (play sequence *f*) she should still go to a psychiatrist, even though the world has changed its strategy. Why? The reason is that Liz is not responding only to what the world did yesterday, but what it did in all its yesterdays, and the impact of its recent change has not yet overpowered the previous good times. Liz is no longer a prisoner of what happened yesterday, but rather a stable person making decisions on the basis of the entire past history of the situation.

What about the world's "decisions" to make Liz happy or unhappy? Liz has said that when she went to a psychiatrist, the world has often treated her well, thereby making her think she did not need a psychiatrist. And when she did not go to a psychiatrist, the world has seemed to treat her badly, thereby making her feel she needed a psychiatrist. Thus she felt that whatever she did, the world would ultimately act to her detriment. On the other hand, Liz is the one who decides how the world is treating her. Thus, there is no objective basis for her claims. To eliminate the painful, idiosyncratic feedback Liz experiences, we eliminate "real" feedback (Liz's perception of how the world is treating her) from Liz's considerations. This is okay because the "real" feedback is imaginary anyway.

The theory of games suggests that the world would use a maximin strategy, which is roughly what Liz was describing: Liz was trying to minimize her pain, and in Liz's eyes the world was trying to maximize it. We apply the same sort of consideration to the actual choices of the world in this repeated problem. Except for the world, the actual choice on any given play is based on the largest total for its two strategies, rather than the smallest total. The world merely looks at what Liz did on the previous play, examines its own running totals, and chooses the strategy associated with the largest running total. The result will be a sequence of choices. Liz uses the world's sequence of choices as the basis for adding and using her own running totals, and the world uses Liz's sequence of choices as the basis for adding and using its own running totals. On the first play of the game, the decisions are made arbitrarily, or put another way, on the basis of how things look to Liz.

FIGURE 19B

How the World's treating me	My total if I go to a psychiatrist	My total if I do *not* go to a psychiatrist	I choose	The World's total if it is *good*	The World's total if it is *bad*	Play sequence
bad	2	4	not to go	1	4	a
bad	4	8	to go	4	6	b
bad	6	12	to go	7	8	c
bad	8	16	to go	10	10	d
good	11	17	to go	13	12	e
good	14	18	to go	16	14	f
good	17	19	to go	19	16	g
good	20	20	to go	22	18	h
good	23	21	not to go	23	22	i
good	26	22	not to go	24	26	j
bad	28	26	not to go	25	30	k
bad	30	30	not to go	26	34	l
bad	32	34	to go	29	36	m

What if the two columns of running totals are of equal value? Then choose either one, it makes no difference. An extended diagram showing both the world's and Liz's choices and totals, and how they interlock for the first 13 play sequences is given in Figure 19B. Providing the reader sticks to the procedure, complete accuracy in the addition is not necessary, since no single mistake is crucial to the outcome. For this reason, it has been suggested that the human brain may sometimes subconsciously use similar procedures to handle complex problems.

What is so special about these two lists of actual choices, one list for Liz, the other for the world? After about 50 plays, count up the number of choices of not going to the psychiatrist. The ratio of these two numbers to each other will be roughly the same as Liz's best mixed-strategy odds in a one-shot version of the game—choose to go three times as often as not go. However, the choices are not mixed up randomly, there are solid blocks of going to a psychiatrist and solid blocks of not going to a psychiatrist. Similarly, the ratio of the world's two choices will be roughly the same as its best mixed-strategy odds, which are fifty-fifty. In other words, we have converted the best one-shot odds into a long-term schedule of choices. Liz could not have a better schedule based on her lowerarchy, because this schedule is insurance against the worst that the world can do. And since no observations of what the world is doing are necessary, the schedule is foolproof.

Liz can make the entire schedule up in two or three hours. Then she merely has to follow it. Each choice on the schedule can represent a single visit to the psychiatrist. So if she usually goes once a week, Liz now has an entire year's schedule.

Liz has resolved her problem.

14 WHAT MIGHT HAVE BEEN

Irving, thirty-five years old, married, with two kids, teaches in the same English department as Jim. Driving home with his wife from a dinner party, they are involved in a serious car crash. His wife, Shirley, is not hurt, but Irving is knocked unconscious. Rushed to the hospital, Irving regains consciousness paralyzed from the neck down. He has suffered damage to his spinal cord. According to the specialist, Irving may be facing a lifetime of paralysis: "Only about ten people out of a hundred spontaneously recover over a period of eight to ten years with rigorous therapy. Recovery means walking with a cane. On the other hand, there is an operation in which the odds of some kind of success are better than fifty-fifty. The detailed breakdown is that forty out of a hundred people completely recover roughly two years after surgery. About fifteen out of a hundred who undergo the operation show significant recovery; in other words, they are ambulatory with crutches or a cane, and can hold down a normal white-collar job. Fully forty-five out of a hundred do not recover; the operation itself seems to bring about permanent damage to the tissues, and the paralysis is a life sentence, not susceptible to therapeutic techniques. So the operation involves a certain risk, and it could come out the wrong way."

Irving's system is still in a state of shock, and no operation can be conducted for at least the next few hours. However, the operation cannot be put off more than three days, thus Irving must decide within the next day or two whether to undergo surgery.

Irving, who is amazingly lucid, tries to discuss the decision with Shirley, who becomes increasingly distraught the more the decision seems to devolve upon her advice. It becomes clear to Irving that Shirley, with the enormous pressures on her, is simply incapable of giving coherent advice. The only other person Irving can trust for clearheaded advice is his father, who has been at the hospital throughout.

The doctors involved in the case refuse either to recommend or discourage the operation. "You have to make this decision yourself, Irving," his doctor says, "I can't tell you what to do; I can just give you the odds. God knows I wish they were better. Remember, if you don't go ahead with the operation, there's always a chance of improvement in medical science—who can say what we'll be able to do in five years' time?"

Irving's father, an intelligent man who has always been healthy, has never had to make any serious medical decisions. From time to time, Irving has consulted with him about various decisions.

"Dad, I don't know what to do. Shirley is so numbed by the accident that I can't get any sense out of her. I can't even talk to her. I think she's in a worse state of shock than I am. I have to make a decision one way or the other within the next two days, and the earlier the better. I'm afraid if I decide to have the operation, I'll chicken out just before I go under the knife."

"There's no problem, son, once you sign the paper authorizing the operation, they'll just keep you on sedatives until it's over."

"Dad, I can't move, I can't sign anything. If I decide to go ahead, you or Shirley will have to sign for me."

"Sure."

"What should I do? I've got two kids. The university will pay my medical bills for a while, I don't even know how long. I can't face the prospect of being permanently incapacitated. I won't be able to teach."

"Son, facts are facts—you may have to face it whether you go through the operation or not."

"What do you mean?"

"Look, the doctor told us, the operation may work, then you won't be crippled—but it may not work. If you don't go through the operation, there's a big chance that you will be paralyzed. If you do go ahead, and it doesn't work, you face a lifetime of paralysis. Irving, you better face it, there's no way you can be sure you won't be paralyzed."

"You mean I had better accept the fact that I'm going to be paralyzed? Has the doctor told you more than he's told me?"

"Irving, believe me, he hasn't told me anything you didn't hear. Yes, there's unfortunately a big chance either way—whether you have the operation or not—that you might be paralyzed."

"So you mean that we should start from that point? Are you sure, absolutely certain? Can we go over it again once more?"

"These are the best doctors available."

"Let's figure that the information the doctors have given us is right. Dad, you've always been able to give good advice whenever I've gotten really stuck—you've got to help me now."

"Irving, finally everybody has to face his own decisions. I'll try to help you make yours, but I can't make this decision for you."

"But don't you see I'm helpless, someone has to help me get at the issues."

"I'll do everything I can, son, please believe me."

"But you said that I've got to start from the premise that I'm going to be paralyzed in any event."

"You may be paralyzed. We don't know. We hope and pray that you won't be."

"Why should we start from that premise? It seems too pessimistic."

"Irving, you've heard what the odds are. The doctor said only ten out of a hundred people who don't have the operation recover, going through some complicated therapy. And if you go through the operation and it doesn't work—he said in nearly half the cases it doesn't work—then you will be permanently paralyzed. You heard what he said, the operation itself may induce further damage. If the operation fails, it's permanent paralysis, period."

"Then why should I go through the operation?"

"It depends on whether you are willing to take the risk, Irving, it's up to you."

"You just can't say it's up to me. I'm asking you, you're my father—I'm helpless. You started from this point of complete pessimism."

"Irving, obviously we both want you to recover fully. Maybe the best chance is if you undergo surgery."

"Dad, you don't understand. It's not possible for me to lie here and throw away the chance of a recovery. If I don't go through the operation . . ."

"Then you have that small chance that maybe . . . but you'll be in bed a lot of years wondering if you're going to recover or not."

"Say that again. What are my odds?"

"Nine to one against you if you don't undergo the surgery."

"And if I have the operation?"

"He said that 40 percent recover fully, 15 percent recover significantly, and the rest don't recover."

"So, it's about half and half."

"A little better than fifty-fifty, but not much."

"But I have nothing to go on, how can I make a decision? I'm not betting on the horses—fifty-fifty I won't be paralyzed if I have the operation; nine out of ten that I will be paralyzed if I don't have the operation. Who knows what will happen? Dad, I can't figure it out this way—it's too confusing. Get a piece of paper and pencil and write down what I tell you."

"Sure, Irving, you want to be logical."

"Got the paper? Write down a four-square grid—yeah, that's right. Put over the top left 'operate' and the top right 'don't operate.' Yeah. On the left, outside the top two squares, 'operation successful,' and 'operation unsuccessful,' along the left, outside the bottom two squares."

"Irving, this isn't a puzzle."

"Dad, I've got to do this step by step, I've got to make some sense of this problem. It's obvious that what I want, what you want, what everybody wants, is to operate and recover. The operation would be the quickest recovery. Put down number 1 in the upper left-hand corner box. Have you done that, let me see."

"Okay, Irving."

"Yeah, now that's best, the highest odds in favor of that, and also what is simply best. Okay, what's second best for me?"

"Irving, there's a small chance of recovery without the operation."

"Yeah, so second best is clearly not to operate if the operation were to fail."

"Obviously, Irving."

"Okay, put down the number 2 in the lower right-hand corner box. These are the obvious things."

"Irving, I don't know what you're doing with this

grid thing, but I have one piece of advice for you—don't think about 'might-have-beens.' If you decide not to have the operation, don't think that it might have been successful, or you'll torture yourself over it."

"Wait a minute, Dad, I'm the one who will have to go through the regrets, not you, not Shirley. I'm the one who's got to live with this problem. Let me think about this. What do you mean? Spell it out."

"Well, it just means you shouldn't think about what might have been."

"What does that mean? Think about it—it means that if I didn't operate, I would have to forget about the operation, pretend it never was an issue. For five, ten, who knows how many years, I'd have to pretend that the operation was not an issue and pretend to forget that it might have cured me. Isn't this more important than anything? What am I going to be doing but lying in bed regretting?"

"Irving, if you want to think that way, you have a choice between two ifs—not operating when the operation might have been successful, or operating and the operation is a failure. I don't think this is a good way to look at it."

"Dad, the first two things we spoke of were plain and straightforward. They were obvious. But now the core of the problem is which of the worst things is really the worst. This is what the decision is really about. Let's go over it again, if . . ."

"Irving, I don't know how you will regret these things. If you don't operate and the operation would have been successful, you've lost a gamble. If you do operate, and the operation is a failure, you've lost another gamble. What are you talking about?"

"Dad, don't you see? It's not that it's a gamble, it's

the regrets, the 'what might have beens' that count here. That's what I'll have to live with. Which regret is worse for me—that's what I have to decide.''

"Irving, be logical, look at the odds, nine to one against you if you don't operate, and less than 50 percent against if you do. Operating is a better gamble. You want to talk about the 'might have beens.' I've never based my life on that.''

"Dad, you're saying that the odds are loaded in favor of the operation, but if I operate the paralysis could be permanent, if I don't operate it may not be permanent, so the odds don't tell everything, don't you see. The crux is the regrets, even with those odds.''

"Son, you mean which of those regrets is going to torture you more?''

"Dad, how could I possibly lie here year after year, and be tormented every day that I didn't have the courage to try. I couldn't deal with that. My children would grow up, I'd never be near my wife, and I would always think that I didn't have the guts to take the best chance that I could. Surely, this is the decision. The worst is to live with regrets that maybe I threw away my chance to recover because I wouldn't take the chance. Dad, doesn't that make sense to you?''

"Sure, Irving.''

"Then you agree with me.''

"I think you'll always wonder about the operation if you don't take the chance.''

"Well, Dad, there may be regrets if I do have the operation, but maybe they'll be less because at least I'll know I tried.''

Irving tells his father that he wants to think it over by himself. He sends for his father a couple of hours later.

"Dad, I've thought about it. I guess we're two different people. When you told me not to look at regrets, I didn't know what advice you were giving— maybe not to operate, maybe to take a chance and do it. But if I'm going to live my life thinking about what might have been, the worst is not taking a chance, so I've decided to go ahead with the operation. I won't be surprised if it fails. It want you to accept this decision; I want you to make Shirley understand it; I want you to get Shirley to sign the release for the operation; and I don't want the decision changed."

Irving's lowerarchy:

4. Irving refuses operation/operation would have been successful
3. Irving agrees to operation/operation unsuccessful
2. Irving refuses operation/operation would have been unsuccessful
1. Irving agrees to operation/operation successful

Irving made the minimax decision based on the lowerarchy in Figure 20. He realized that the crucial issue was a choice between the two most painful "if's": operating, and the operation was a failure; and not operating, and the operation would have been a success. Irving decided that he had to make his decision based on which of these two situations was worse. This is a minimax decision, a pure strategy minimax. Of course the conflict map is circular, and in strict game theoretic terms calls for randomization. Irving had to make a decision he could live with, and he could never have lived with a decision based on randomization. Thus he had to use the pure strategy minimax. This was a logical decision based entirely on what is often thought of as an illogical idea, thinking about "what might have been."

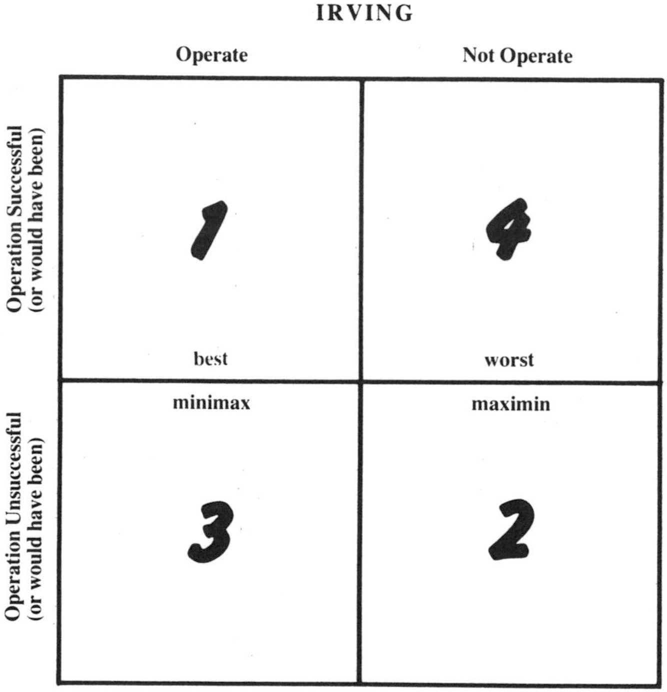

IRVING

FIGURE 20

Let us suppose that instead of Irving, his seven-year-old child was paralyzed in a similar accident. Assume that the medical facts are the same. Would this change the decision? Maybe yes, maybe no. Obviously the outcomes ranked 1 and 2 on the conflict map would remain the same. Operating when the operation is successful is still "best," and not operating when the operation is a failure is still second best. The question focuses on the two remaining outcomes. Would they be ranked the same? The child is very young. Maybe Irving and Shirley would figure it better to bank on the progress of medical science rather than risk permanent damage in the operation. Yet, the parents still have to look at regret. The child may always accuse them of not taking the risk, and they may accuse themselves of this as well, of damning their child to a lifetime of paralysis. We don't know how Irving and Shirley would feel in this hypothetical case.

How would you feel?

15 SHOWDOWN WITH COLONEL BLOTTO

Liz has worked for several years as a volunteer at the County Museum in the membership department and recently has been elected to the membership committee. The strongest personality on this committee, who has been effectively in control for many years, is an ex-military officer and retired city manager, Colonel Blotto. While he is away on an extended vacation there is a meeting of the committee at which Liz proposes two first-rate ideas for an extended membership drive. The committee accepts Liz's suggestions and she is put in charge of both projects.

The proposals were enthusiastically supported by a younger group on the committee who dislike Colonel Blotto and resent his rigid and authoritarian style. On his part, Blotto resents younger people and always belittles their ideas. Of course, all the members of the committee are in it for its own rewards; no pay is involved. Many of the members also want Blotto eased out and a new spirit and energy brought to the committee.

Liz's proposals would change the membership structure of the museum, which now consists mostly of older people. The basic idea is to bring in a whole range of younger people. Proposal A is joint membership for married couples with small children; Proposal B is to bring in college students.

Colonel Blotto returns and is upset that the most adventuresome proposals in the history of the membership drive are not under his control, as has been customary. Not knowing what to do about this, but being a very influential man in the museum and in the community, he has lunch with the chairman of the membership committee. He makes no specific attacks on either proposal; in fact he defends the projects in principle, but raises doubts in the chairman's mind about Liz's competence to handle them.

Realizing that a large sum of money is involved in this extensive membership drive, the chairman of the membership committee becomes alarmed. He decides it would be prudent to call a special meeting of the committee to re-examine Liz's proposals for handling the projects. (This way he figures he is covered if either or both the projects fail.) He tells Colonel Blotto, "I'm going to call a special meeting; you can raise any objections you want. But remember, Liz has always been extraordinarily competent, and I wouldn't be surprised if she could adequately defend her position." Blotto reacts by saying, "We'll see at the meeting."

Liz is informed by phone that the meeting is to be held in several days' time; she is advised that there are some doubts about her ability to handle both or either of these complex proposals, and told that the committee will go over her plans to implement them. Liz immediately phones a couple of younger committee members with whom she is friendly to tell them what's happened, and also to see if she can find out more of what Blotto is up to. Her friends are incensed at Blotto's meddling, which is giving Liz extra work and everyone more meetings—"a terrible waste of time." They suggest to Liz that she should prepare to squash Blotto utterly, thereby diminishing his influence on the committee.

Blotto hears of this scheming. He then realizes that he is going to have to stand and deliver. To win his case he will have to present objections to the manner in which Liz handles her proposals by throwing detailed doubts on her administrative ability. The proposals are only known in broad outline; however, for implementation, it is necessary to flesh them out in full detail. Blotto realizes that the only way he will be able to win in the showdown ("Carry the day," as he puts it) is by being able to fault Liz on her detailed analysis. This will require working out detailed evaluations of each of the proposals. He has only a few days to do this. And as Blotto has rather limited energy, he can only do a very thorough and detailed analysis of one project, or a less than competent analysis of both. Liz, who works during the day, has time only to do a thorough defense of one or the other project, or a less than adequate defense of both.

Liz talks the problem over with Jim. She explains that the meeting has been called a few days hence, when everything will be decided, and explains that Blotto will raise objections to her handling of either project A, or B, or both. But she doesn't know which one he will attack. She does know, however, that she could do a detailed defense of either proposal. She knows that if she does a thorough defense of one and Blotto attacks it, no matter how effectively, she can beat him back point for point. She has the time to do this for one project only. Likewise, Liz figures that Colonel Blotto is in a similar position: He has only the energy and time to review one project in detail. The question is, will Blotto attack her on the same project for which she has prepared an adequate defense? If this happens, Liz retains control of both proposals while Blotto's objections will be thoroughly smashed. This is the best for Liz, bottom line on her lowerarchy.

Jim says to her,

"Okay, it's best for you if you match, but you don't know which he is going to attack, proposal A or B. We're pretty safe in figuring that he doesn't have the energy or time to do a thorough analysis of both. Apparently, from what you've found out from the other members of the committee, this meeting was as much a surprise to him as to you. He was grumbling with the chairman, and suddenly he had a meeting on his hands. So he won't have time to do a thorough job on both. He could, of course, do a partial analysis of both. Look, the thing you want most is to hang onto both proposals, and that will only happen if you collide on the same proposal, if you match each other. Right?''

''True.''

''Well, would you mind giving up one of the proposals? After all, you're a volunteer, this is taking a lot of time and I'd like to see you more often. Why not give up one project?''

''If I have to give up one, I suppose I care least about project B. It's an interesting proposal; I'd like to get involved with it, but the other one, the one for young married couples, is more useful. I'd be bringing into the museum a very stable group who maybe could contribute a lot of energy and ideas, whereas students come and go. Since I only have time to prepare a detailed defense of one project, suppose I prepare a detailed defense of A and Blotto attacks me on B. I would be stymied. Oh, I could say a few things, but he'd be able to fault me, I'm sure. Blotto would win on that one, taking B away from me, which the committee would have to give him control of. Having demonstrated that he could handle it, he'd get it.''

''If Blotto got project B, would he be satisfied? Couldn't he go running back to the chairman of the committee to try to grab A as well?''

''No. He'd be too late. I would have prepared a thorough defense of A, and before I'd allow the meeting to adjourn I'd give a report on it. They'd have to let me keep it. Next year we might go through this again, but for this year I'd have no more trouble with him. However, if I can successfully defend the one he attacks, he'd be defeated. If I could do it ably and competently, the other members of the committee who are wavering about Blotto's situation, and who know that others are unhappy with him, might turn against him. There is the possibility that all his power would be reduced to zero, and lots of younger people would be given these jobs to do. We'd have a whole new spirit and energy on the committee."

"But Liz, this can only happen if Blotto attacks you on the same proposal for which you have prepared an adequate defense. If you miss—if he attacks A and you defend B, or he attacks B and you defend A—he'll get whichever one he attacks. Right?"

"Unfortunately, you're right."

"Well, you've decided that keeping both is best, bottom line on your lowerarchy—matching him, in other words. If you defend A and he attacks B, this is second best for you because you at least hold onto the project you prefer."

"That's right."

"So third on your list is: If you defend B when he attacks A, he wins the project you most want—project A—and you will be stuck with B."

"Again, unfortunately, yes."

"Well, what happens, Liz, if you were to prepare a defense for both projects? I know it can't be adequate, but suppose you were to prepare a sketchy defense of both in the next few days and he attacks you thoroughly on A or B?"

"Oh, if I did that and he attacked me sharply on either one, I'd lose. In his way, Blotto's very smart and experienced and I think that some doubt would be raised as to my competence to handle both proposals."

"You could lose both?"

"It could happen. He really resents anyone doing anything—other than himself—unless he is fully in charge."

"What happens if he attacks you in a sketchy way on both but you thoroughly answer his objections on one, completely crushing him on that one?"

"Well, he'd be reeling from my defense of one, and I could just say I hadn't had time to prepare a thorough defense on the other. In fact, I probably wouldn't have to even say that. The committee would assume it. He would just look like a meddling old fool. Every member of the committee would figure that his attack on the second proposal would be as shallow as his attack on the first. I have no doubt I'd retain both projects and he'd be thoroughly crushed."

"What if both of you make sketchy analyses and reports on both projects?"

"I don't think Blotto would do it. But if he did and I did, I guess I'd limp through with both projects just because enough members of the committee would back me. No, I suppose it's possible they would give Blotto project B. Yeah, I'd better figure that's what would happen."

Liz's lowerarchy (Figure 21):

4. Liz defends A and B/Blotto attacks A
4. Liz defends A and B/Blotto attacks B
3. Liz defends B/Blotto attacks A
2. Liz defends A and B/Blotto attacks A and B
2. Liz defends A/Blotto attacks B
1. Liz defends A/Blotto attacks A
1. Liz defends B/Blotto attacks B
1. Liz defends A/Blotto attacks A and B
1. Liz defends B/Blotto attacks A and B

Blotto realizes that the younger people on the committee will support Liz unless he makes very adequate prepara-

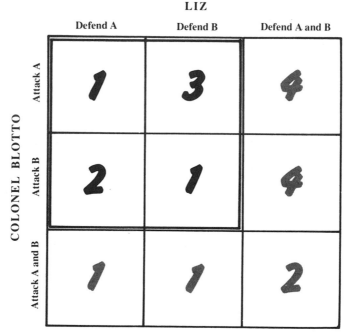

LIZ

NOTE: Each number in the first or second column is smaller than the corresponding number in the right-hand column. So, the right-hand column is rejected by Liz. Similarly, each number in the top two rows is bigger than or equal to the corresponding number in the bottom row. So, the bottom row is rejected by Colonel Blotto.

FIGURE 21

tion; if his proposals aren't to the point, he cannot maintain his superior position on the committee. He knows that he is threatened with being eased into the background, and that the younger people want to take the committee over. He has to work out exactly what he intends to do. Ideally, best for him would be to get both projects under his control. He talks to his staff (his wife) at headquarters (his home).

"My dear, this is a very interesting tactical problem."

"Yes, Colonel, but how can you get both of these projects?"

"Only by discrediting Liz thoroughly."

"How?"

"By going for broke—a detailed analysis of both proposals that raises serious questions about Liz's competence."

"But Colonel, you don't have time in the next few days to do a thorough analysis of both projects."

"I'll take my chances and get her on the run."

"You mean, do a second-rate job on your objections?"

"Well . . . you know they're trying to ease me out. I didn't mean for this situation to occur—it wasn't in my battle plan. I came back from vacation, and found out about this sneak attack . . ."

"Colonel, stick to the point! If you do a second-rate attack on both of these proposals, Liz is going to beat the pants off you and you know it. Even if her defense is second rate, the committee likes her best. You wouldn't get any more than project B. So, which project will you attack?"

"I'll attack the Big One, project A."

"But what happens if she defends the Big One? You told me earlier today you thought Liz was doing an outstanding job on both proposals. You just don't like her attitude."

"Right. The clever strategy, of course, would be to go for project B and make her look like an idiot. I'd rather have project A, but B is serviceable."

"But what if she expects you to do this and prepares an adequate defense of project B?"

"Well, it depends on how she thinks. She's shrewd. Maybe she thinks that I'll go after A, so she'll defend it. Anticipating this, I go after B; but she may figure this and defend B instead of A; so maybe I ought to go for A after all."

"But what if she goes through the same thinking and defends A?"

"Well, what's the answer to that? She and I both know that project A, the young marrieds project, is the Big One. We both want it. I don't really want proposal B, but I want something. She knows this, and if she figures I'll go after A, she'll defend it, and the committee being stacked against me, I'll lose everything and be laughed at. On the other hand, knowing this she may figure I'll attack her on project B, the college-student project. But if she figures that . . . she's probably ready to give up proposal B. After all, project A is the more important one."

"But Colonel, what if she's not? What if she thinks you will think that?—she'll defend project B and you lose out completely."

"What do I do?"

"Colonel, I don't know. Maybe you should give up."

"No!"

Blotto's lowerarchy (Figure 22):

4.   Blotto attacks A/Liz defends A
4.   Blotto attacks B/Liz defends B
4.   Blotto attacks A and B/Liz defends A
4.   Blotto attacks A and B/Liz defends B
3.   Blotto attacks B/Liz defends A
3.   Blotto attacks A and B/Liz defends A and B
2.   Blotto attacks A/Liz defends B
1.   Blotto attacks A/Liz defends A and B
1.   Blotto attacks B/Liz defends A and B

Blotto's lowerarchy is virtually the opposite of Liz's. As is clear, this is a pure conflict situation between the two, based on the principle that both cannot occupy the same space at the same time. (Many power grabs in politics, business, charities, and public service are of this type.) Blotto

COLONEL BLOTTO

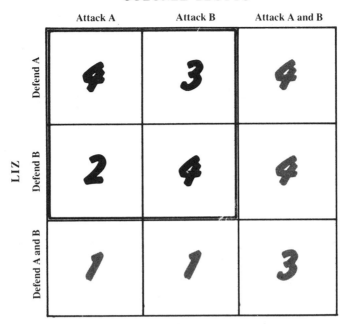

FIGURE 22

does not yet know what to do except that he has rejected the option of preparing a generalized attack on both of Liz's proposals. He rejects this, realizing that he would lose in every case except when Liz also presents a generalized defense of both A and B, which he cannot count on, and which at best would only give him the college-student project. Thus Blotto can easily reduce his options down to a detailed attack on either A or B, but he finds it not so easy to decide on which one (Figure 23).

118

COLONEL BLOTTO

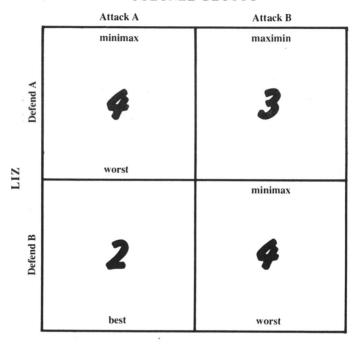

FIGURE 23

He knows that the big prize is project A, and he realizes that Liz, knowing this, will probably defend it. But he doesn't know for sure. Liz would seem to be extremely unwise defending project B rather then A, because A is the more important project. But Blotto also knows that if Liz can figure out which project he will attack, she merely needs to defend it to eliminate him. Since Blotto figures Liz is most likely to defend A, he would be shrewd—given the shortness of the time—to marshal all his resources and make a decisive attack on project B, which, if successful, would reinstate him in the eyes of the committee. But then, reflecting further, he realizes that Liz may anticipate he will do just that, and defend project B. Blotto is nonplussed.

With his mind in turmoil, Blotto finally faces the fact that a decision has to be made fast—and one that Liz cannot possibly anticipate. Liz is a very astute person who may see his weaknesses (his wife has often warned him about his vanity). To prevent himself from leaping at the Big One, and losing all, Blotto decides to surrender his own reasoning power by choosing randomly. To play it slightly safe, he adjusts the odds in favor of project B, making them two-to-one on B, and thus one-to-two on A. Flipping a coin won't do because that gives odds of fifty-fifty. (Even the best two out of three flips still gives fifty-fifty, since each flip is fifty-fifty.) He has to find another device to randomize, thereby preventing Liz from figuring out his choice on the basis of his personality. Requiring a method loaded in favor of project B, he tells his staff to take three matches, break off the end of one, mix them up thoroughly so she doesn't know which they are. If he draws the short one, he chooses project A; otherwise he chooses B.

Liz is in the same pickle as Colonel Blotto and rejects the option of defending both A and B, since with this option she can only win if the Colonel attacks as weakly as she defends—in other words, if he attacks both A and B. But she is aware that Blotto is not stupid, so it is in absolute certainty that Blotto won't attack both. This leaves Liz with the options of doing a first-rate report in defense of either project A or project B (Figure 24).

Liz realizes that A is the more important project and thus she is most tempted to defend A and has least to lose by doing so. But the temptation to match Blotto and beat him at his own game is overpowering. Liz decides she is willing to run a small risk to defend B. Furthermore, she is afraid Blotto will figure her for A and therefore attack B, taking it away from her. To be certain Blotto cannot figure out her choice Liz must be unable to figure it out herself and to randomize with the odds favoring project A. But how to make her choice inscrutable and safe at the same time? Liz goes out

LIZ

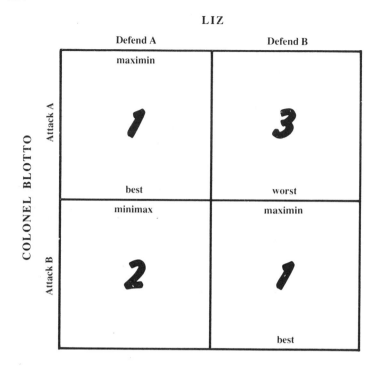

FIGURE 24

for a walk to think about the problem, leaving Jim at home. When she comes back she finds him watching a hockey game on television. Suddenly remembering that a hockey game is divided into three periods, Liz decides to defend A if the game is in the first two periods, or defend B if the game is in the last period. So she asks, "Jim, which period is the game in?"

Given his unfavorable position, Blotto is forced to put his heaviest odds on an attack of project B, even though he would rather have project A, since he feels it most likely that Liz will defend project A.

Liz and Colonel Blotto acted rationally by surrendering their choice to chance. The risk each takes is completely calculated. Even playing at his best, Blotto can completely lose, and at the most, he can only win one project. Liz, if she plays well, can at worst lose project A.

Blotto's mistake was to get involved in this game in the first place.

Both Liz and Blotto are using the safest possible gambles—true, calculated risks. Consider Figure 24. Liz uses 2/3 on A and 1/3 on B. If Blotto attacks A, Liz can expect $2/3(1) + 1/3(3) = 5/3$. If Blotto attacks B, Liz can expect $2/3(2) + 1/3(1) = 5/3$. Whatever Blotto does, Liz can expect the same amount. Furthermore, she only runs an 11 percent chance of losing project A, and stands an 89 percent chance of either matching Blotto, and thus thoroughly trouncing him, or only losing project B. Blotto, on the other hand, by using 1/3 on A and 2/3 on B can hold Liz to this expectation: Suppose Liz defends A; Blotto holds her to $1/3(3) + 2/3(1) = 5/3$. Suppose Liz defends B; Blotto holds her to $1/3(3) + 2/3(1) = 5/3$. Thus whatever Liz does, Blotto can hold her to the same expectation. This is the best that Blotto can do.

Perhaps Blotto should have applied the Jimmy-the-Greek odds system to his lowerarchy to include his intensity of fear over matching Liz—in which case Blotto loses everything. However, Blotto felt that the simple lowerarchy was sufficient to guide his actions: "I am not writing a bunch of scientific baloney on how I feel," he explained to his wife.

16 COLONEL BLOTTO'S RETREAT

Both Liz and Colonel Blotto have apparently prepared their reports as we find them at the emergency meeting of the membership committee. The chairman has just called the meeting to order. All eyes turn to Colonel Blotto, everyone expecting his attack. Blotto addresses the committee:

"My fellow members, some days ago a suggestion was made to me that there were serious defects in Liz's plans for the two membership drive proposals. I mentioned the fact to our honorable chairman, who thought it necessary to call an emergency meeting of the committee to discuss the matter. Naturally, having known Liz for several years, and having a great respect for her intelligence and competence, I seriously doubted that anything could be wrong with projects to which she devoted her attention. However, I thought it my duty to investigate these allegations personally, confident that they would not stand the light of day. I have looked at both of Liz's proposals very carefully, and I have talked, in strictest confidence, with supporting members of the community, as well as persons active in the museum. I have gone over in exhaustive detail exactly what Liz has done, and what remains to be done, on both proposals. (Liz smiles, expecting a second-rate attack by Blotto on both proposals.) I want to report now, because I feel it imperative to do so (hushed silence in the room as everyone waits for the shoe to fall) that not only have I found nothing on which Liz can be faulted, but that I have only the highest admiration for Liz's handling of these proposals. Liz is, indeed, doing an absolutely first-rate job, deserves our full support, and should be commended by all of us. As for those who whispered the allegations against her, they know who they are, and must secretly bear their shame. I will not divide this committee by pointing the finger. I beg Liz, should she ever learn from another source who these persons are, to forgive them. I, for one, offer my complete and unqualified support, and extend an offer to help Liz in any way I can in whatever difficulties may arise in the future. In any case, the scurrilous rumor which went around that Liz was not doing a good job should be thoroughly quashed by all of us. Thank you.''

There is a puzzled silence in the committee. Everyone looks at each other suspiciously. Is it true that other persons, not Blotto, started the trouble in the first place? Liz sits stunned. Her notebook, with a careful defense of one of the projects, sits closed in front of her. The chairman feels compelled to say something: "Well, I guess it's unanimous then. We're all agreed that Liz is doing a fine job on both of these projects, and I think it very generous of Colonel Blotto to offer his assistance. Will that be of help to you, Liz?''

What can she say? She can't very well say no at this point. She is forced to publicly and graciously accede, whatever her private feelings.

Explanation: Blotto, in finding that the odds were overwhelmingly against him (he only had about an 11 percent chance of winning project A), had to turn the tables and attempt to impose an image of himself as a man of generosity, intelligence, high energy, and responsibility. Although he made it sound as if he had researched both projects, in fact, once he calculated the odds, he researched neither. By praising Liz he made a generous strategic withdrawal which positively enhanced his public image in the eyes of the committee. He also showed that he did not want to divide the committee. Rather, he wanted it to move ahead, not split into factions by petty squabbling, but as one voice. In this way he enhanced his image as a peacemaker and an elder statesman.

How was game theory helpful to Blotto? First, it helped him to structure the problem and showed him that he could effectively attack only one proposal. He certainly would have lost had he attacked both. Second, in calculating the risks, Blotto found them too great; game theory told him not to play this game. Realizing that he had suckered himself into a losing game—the odds nine to one against him—Blotto decided on a strategy of generous withdrawal. Liz was totally unprepared for this new game. Now, under the guise of cooperation, Blotto can gradually gather back his power over a period of time.

Since Liz has said she will be happy to have him help her, he merely has to go ahead and run the projects as if they were his own anyway. As he has far more connections in the community than Liz, he merely has to phone around, say he is helping her, and if there are any questions, contact him; he will pass the information along to Liz. He has put Liz in a position where she will have to raise the issue that he is interfering. She will have to attack the elder statesman who, to all the world, is helpful and interested in avoiding petty squabbling and ego politics. If the projects fail in any way, Liz takes the blame, since they are both her projects. ("They were, of course, Liz's projects, but I did what I could to help her.")

Blotto's objective is to get cooperation on his own terms. Furthermore, he is in a position to withdraw from this new game at any time, for any number of excuses, if he sees that the projects will not be successful. Liz does not have this option.

Is there any way Blotto could have been shot down at the committee meeting? Liz did not have to reply as she did.

She could have said, "The proposals were originally mine. There are very experienced people on the committee and in the museum who examined these proposals in great detail and who agreed during Colonel Blotto's absence to proceed. Colonel Blotto returns, throws the whole project into doubt, and we have had to run and have a special committee meeting. I've wasted a week preparing a defense of my proposals, which he has analyzed and found perfectly okay, and we are now back where we started. Colonel Blotto conveys a public image of himself as a generous, helpful person. In fact, he is just a silly old patronizing timewaster. I don't want to work with him, I don't want him on this project, and I don't want him on the committee." There is a reason why Liz does not say this: It's the wrong time. Liz could have said this had she defeated Blotto in the showdown. But there was no showdown. Thus such a speech would be unseemly—divisive, petty, and egotistical. The chairman and many committee members would be outraged. Blotto offers unity, Liz divisiveness. There is nothing to be gained by Liz forcing the crisis now. She would look like a person more interested in ego gratification than the welfare of the museum and the community it serves.

But suppose the reverse situation had occurred, with Blotto originating projects A and B and Liz wanting to seize control of them for herself. Were Liz to attack, the odds would be equally weighted against her as they were for Blotto. If Liz, after calculating the odds, were to attempt to employ the strategy of "Colonel Blotto's Retreat" to reinstate herself in the eyes of the committee, the ploy would fail. Why? Liz, due to her public image of youth and inexperience, would look more like a failed opportunist than a cooperative and public-spirited young woman.

17 FLASHPOINT!

In earlier examples, our concern is with situations where the decision maker has a clear choice—a pure strategy—on the conflict map. The crises always contain a flashpoint, a moment where something has to be done to implement the decision, even when inaction is a final decision. Since the decision is based entirely on the person's own lowerarchy, his or her own values, these determine the course of action. Once the lowerarchy is laid out, the decision becomes clear and inevitable.

In "V.D.—Light and Easy" and "Showdown with Colonel Blotto," examples with circular conflict maps (where the person is torn between alternatives), there again is a flashpoint. A decision has to be made.

In the V.D. example, May is torn between telling or not telling Ken whether she has slept with someone else or not. Her lowerarchy shows her feelings in tremendous detail. However, faced with the palpable presence of Ken across the table, can May keep her randomized course of action? Can she even use her minimax pure strategy, her safest course—keeping her mouth shut? Remember, she fears Ken might set her up to feel guilty. She wants to tell him—she likes the idea of being free, of not being tied down.

126

What binds her to the result of a randomization strategy? Nothing. This is why she defers the decision until the last possible second, to keep herself from having "second" thoughts. So she hopes. In any case, game theory shows the need for simultaneous disclosure.

May deals with the situation as if she cannot talk openly to Ken. Affairs should not be games. People should talk to each other honestly. Children should eat their spinach. The situation should not be one of pure conflict between May and Ken, but in May's mind it is because she has another problem—fear of being trapped.

In "Showdown with Colonel Blotto," the situation inherently involves commitment. To do a first-rate report is a time-consuming job. Once either Blotto or Liz starts his or her report there's no time to change; it's too late. Had Liz somehow found out that Blotto was preparing his report on the opposite project, Liz would still be stuck with the one she started. She would have been committed. If she changes in midcourse, her reports would be second rate. The lack of time itself often imposes a decision, like it or not. In "How's the World Treating You?" Liz commits herself to a schedule by arranging it with her psychiatrist in advance. His crowded calendar will keep her from going when she is not scheduled to go, and the fact that she goes only for a few months at a time keeps her from souring on it—keeps her temporarily committed. Although there is no actual flashpoint, in "Dividing the Divisive Heirloom," the intolerable arguments eventually force a solution.

Commitment is inherent in the nature of the strategies of these situations because they involve flashpoints where some action has to be taken.

Often there is no flashpoint, for example, when a person decides he or she is unhappy with his or her present life style or occupation. The desire to change is here, as is the fear of change. The person is safe and comfortable. But he or she wants more—money, fame, glamour, the inner satisfaction of pushing to the limits, or quite the opposite, an escape from material ambition and a shift to public service or a return to the land. Even the knowledge that one has not really tested oneself in the crucible can in itself be a source of intense dissatisfaction—a smouldering crisis without a burst of flame. Time slips by and one's life can be wasted in one's own eyes. Now commitment to the new course becomes a dragged-out crisis, one which never comes to a head.

The solution is to create a flashpoint where a decision has to be made.

Jim, a college teacher, has always wanted to be a writer. He has written a number of short stories, none of which has been published although he has shown them to his friends, who encourage him to continue writing in his spare time. He has also shown his writings to his colleagues in the English department, who encourage him to continue writing, as they do—on their days off from teaching.

Jim sets out on an ambitious project, a novel. One summer he writes three chapters. Enthusiastic but cautious, Jim knows that to continue is a big commitment. He hesitates to proceed further, feeling that he ''can't do any more,'' as he says to Liz. She presses him to admit his fear that the rest of the book may be no good. Liz insists the three chapters are terrific and asks Jim to write out a brief outline for the rest of the book. She hounds him for a month and finally he does it.

Liz takes the three chapters and the outline to a friend, Ellen, a literary agent, to test her reaction. Ellen is enthusiastic, gets in touch with Jim, and explains that she'd like to submit his proposal to publishers. Jim jumps at the chance.

Ellen sends the proposal around to a few publishers, who reject it saying, ''He's an unknown quantity, we'd have to see more before we could commit ourselves.'' Finally, she sends it to a publisher who says ''Yes, this is very promising. We like what we've read and we want to meet Jim and talk it over with him.''

They have lunch, discuss the novel, and get on very

well. At the end of the lunch, Jim says to the editor, "Well, what's the next stage?" He replies, "I guess it's down on the mat with Ellen over the terms of a contract—the best two falls out of three."

Time goes by and Ellen phones to say she has an option contract which includes a small advance. She explains that he can't expect more on a first novel since he is an unknown and the novel is far from finished. "Take the offer," Ellen says.

The contract arrives and sits on the coffee table for a few days. Jim can't bring himself to sign it—he is having second thoughts. He realizes that if he signs the contract he is committed to writing the book. Jim fears if he fails to produce the book, or the book is no good, he will be finished in his new career. "Maybe there's a blacklist of writers who take advances and fail to produce the goods," Jim explains to Liz. He has had fantasies about being a famous writer—fame in his dreams is more satisfying than the reality of possible failure.

Jim discusses his lowerarchy with Liz:

"Look, Liz, you know how I feel. Obviously what I want most, bottom line on my lowerarchy, is to sign the contract, write the book, and be a success. But what scares me, though, is to sign the contract and maybe not produce the book. I don't know if I can even write the thing. Or if I write it, the publisher may not like it, may turn it down. Even if they publish it, the book might be blasted apart by the critics. Or it might be so lousy it won't even get reviewed. By signing the contract I'll have a deadline to meet. I ought to write my first book at my own pace. So if I sign and the book is no good—that terrifies me. On the other hand, if I don't sign and the book is no good, that's second best. Maybe what I should do is not sign and just go on working on the book at my own pace."

"Jim, you know if you don't sign the contract you'll

never finish the book. You need a deadline to get you writing again. You say that it's worse for you to sign the contract and have the book be a failure—not finished, not published, not liked, not something. I understand your feelings. But remember, success doesn't come overnight, and you're leaving out something really important: What if you don't sign and you write the book, and it's good—it would have been a success. You know, with editors in publishing houses it's hit or miss. Sometimes they turn down terrific books, even masterpieces, and maybe best sellers. They turn them down, they don't know. If you don't sign you might write a good book and never get it published. Right now you've got an editor who believes in you. What if you don't sign and the book would have been a success?''

''Yeah, every time I produce a good chapter it would be like kicking myself because I hadn't signed the contract. I guess that's even worse than signing the contract and the book turning out to be no good. That's really the worst for me. And, although people in publishing say a really good novel will find a publisher—and command a larger advance—I don't really believe it. No matter, I need the deadline to work against.''

''Okay, now we've got your lowerarchy'' (Figure 25):

4.   Jim doesn't sign/book would have been successful
3.   Jim signs/book a failure
2.   Jim doesn't sign/book would have been a failure
1.   Jim signs/book successful

Jim examines his conflict map and sees the circularity. The conflict map calls for him to flip a coin, which he does—heads sign, tails not sign.

JIM

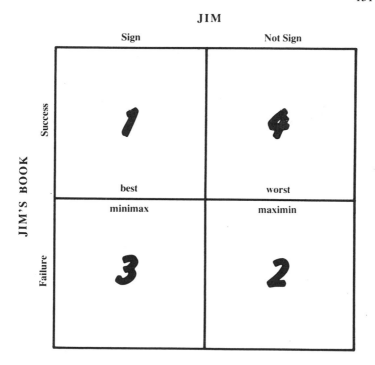

FIGURE 25

The coin lands tails.

"To hell with the coin, I'm going to sign." Signing is Jim's pure strategy minimax—his safest course.

But the contract remains unsigned on the coffee table. Jim keeps reading it until he knows it by heart. Every time he starts to carry out his decision to sign, he hesitates—"I can always sign tomorrow." Throughout Liz has been the enthusiast; her enthusiasm has carried Jim this far. She discusses his dallying with him and they realize the problem—no flashpoint where Jim has to sign.

Finally, Jim says to Liz, "I'd like a party Saturday night—we'll invite all of our best friends."

"Why?"

"I'm going to celebrate the contract. I think that by announcing it to all our friends I'm imposing a sanction on myself of their disapproval if I don't sign. Then I know I'll sign."

The party takes place. A few friends inspect the contract. Somebody says, "Hey, Jim, you haven't signed this thing!" Jim says, "Good God, I forgot." And he signs.

For Jim to back down on signing in front of all his friends would be to lose face. He has imposed a sanction on himself.

Thus, sometimes when someone boasts, he does so not because he is a braggard who can't keep his mouth shut, but because he needs disapproval against backing down. Boasting is one way of committing oneself.

Months go by. Jim is busy teaching—faculty politics, committee meetings, student advising, and a heavy teaching load. Instead of buckling down on weekends and the odd periods between teaching, Jim fritters away his time. Liz realizes that Jim is getting more and more evasive. Every time she broaches the book he becomes touchy.

Months pass. Jim has written one more chapter, and is nowhere near the schedule he set for himself. He only wrote the one chapter because his friends kept pestering, "How's the book coming?" To which he could honestly reply, "Well, it's coming along slowly, but coming."

Finally Liz gets mad. "Jim, you're not producing. In a few years you're going to hate yourself, saying, 'I let the one chance I had to get out of the rut slip through my fingers.' You're simply not producing; you've always got an excuse—a new course to prepare, a faculty meeting. But what's really bothering you is you're scared, afraid to put yourself on the line and do what you think your character calls for—to be a writer. I believe you can write it. All of your friends believe you can write it. Your agent believes you can write it. The publisher believes you can write it. You're scared. And it's paralyzing you. You must force another crisis, make the decision to buckle down and write. Quit your job. Then with your back to the wall, you'll be committed, you'll have to do the book. You can always get another job, maybe not as good as the one you have now, maybe it'll take a year to get another job, but you can always get another job. But you may never have another chance like this."

Jim reluctantly agrees and, to Liz's surprise, admits that he has already thought about quitting teaching, and has worked out his lowerarchy for this decision (Figure 26):

5. Jim doesn't quit/would have been a very successful writer
5. Jim doesn't quit/would have been a moderately successful writer
5. Jim doesn't quit/would have had little success as a writer
5. Jim doesn't quit/would have been a failure as a writer
4. Jim quits job/is a failure as a writer
3. Jim quits job/has little success as a writer
2. Jim quits job/has moderate success as a writer
1. Jim quits job/is a very successful writer

Though Jim's clear preference is to quit, every time he tries to act he defers the issue. "That's why I've been so difficult lately. The only way I'm going to write is to put a painful sanction on my head to produce," he admits to Liz. But now he has to commit himself to the commitment, to the sanction. However, there is no flashpoint, no definite moment when he has to quit the job.

At the end of the fall semester, Jim arranges a dinner party. He only invites his colleagues in the English depart-

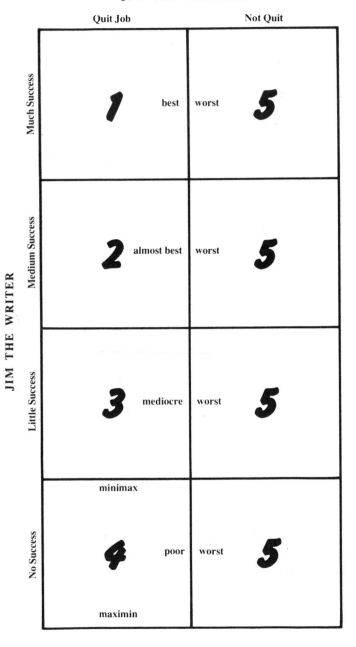

FIGURE 26

ment and their wives. Everyone is seated around the table when Jim stands up and says, ''I'd like to propose a toast to Dr. Rand, our distinguished chairman. Bill, it's been fun, but I quit. As Shaw said, 'Those who can, do; those who can't, teach.' I'm going to see if I can!''

Explanation: Jim burned his bridges.

18 THE PRISONER'S DILEMMA—MORONS DO BETTER THAN LOGICIANS

People are sometimes tempted by unexpected situations which more often than not they resist, if only out of fear of being caught. Recently the papers carried the story of two squad-car police officers stopping an expensive car on a minor infringement. The car turned out to be driven by a well-known underworld narcotics figure, whom the police recognized. But they had no obvious basis for an arrest. However, the police spotted a gun—unregistered. As they were about to take the man into custody, he said, "Wait a minute, let me show you something." He opened the car trunk and the two hard-working officers stared at over $100,000 in small bills. "You may as well take the cash and let me go because you both know that my lawyer will get me off anyway."

What stopped the cops, the newspapers reported, from taking the bribe was that they were new partners and neither knew if the other might be a special investigator. Result: Some police forces around the country are breaking up the buddy system.

Although rare for police to be caught in this dilemma, they often use it to catch crooks.

Two men attempt to pawn some jewelry. Checking

the police list, the pawnbroker determines that the jewelry is stolen and notifies the police; the two men are arrested for possession of stolen goods. They are placed in separate cells and interrogated individually.

The police genuinely believe that the suspects are the robbers, but have no proof. Only a confession from one, the other, or both will enable the police to obtain a conviction. Each prisoner is offered a chance by the D.A. to plea bargain. Each is separately told, ''If you sign a confession that you and your partner committed the robbery, you'll get off by becoming a witness for the prosecution. You'll go scot free, but your buddy takes the rap, and gets the maximum sentence. We'll release you; you'll just have to appear at the trial as state's witness.''

When either suspect asks, ''What if that crook confesses too?'' the interrogating officer replies, ''If you both confess, you both go to the slammer, but for a lot less than the maximum. If you keep quiet and your buddy signs, you'll get the maximum.''

''But since I'm innocent, why should I confess?''

''If you want to stick to that story—good luck, dummy. When your buddy confesses, you'll do ten years and he'll go free. Maybe he won't confess. And if you don't—we only have each of you for possession of stolen goods—you'll both get six months at best. Your only chance of walking out of here scot free is to sign, and remember, your buddy can also sign any time, not just today, but any time.''

The lowerarchy for each suspect is shown in Figure 27.

Each box contains two numbers separated by a broken line; on the left is what suspect A gets, on the right is what suspect B gets.

Each suspect is only interested in saving his own skin. If A confesses, each outcome is better for himself than not confessing. In other words, regardless of B's choice, if A confesses and B also confesses, A gets two years. This is bet-

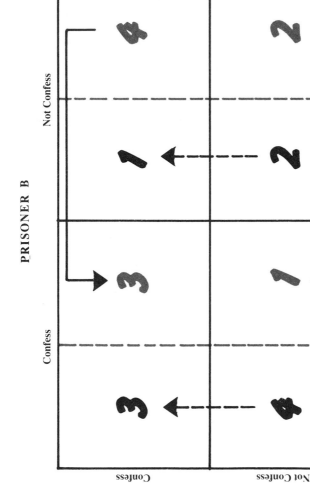

PRISONER B

(column headers) Confess | Not Confess

(row labels, PRISONER A) Confess | Not Confess

	3	1
	3	1
	4	2
	4	2

NOTE: The arrows show that one prisoner always has an incentive to change strategies except when both prisoners confess.

LEGEND: 1 = scot free.   3 = two years in prison
             2 = six months in prison   4 = ten years in prison

FIGURE 27

ter than ten years, which is what A will get if B confesses and he does not. And if A confesses and B does not, A's outcome is better than if neither confess—scot free is better than time in jail.

Each suspect figures that his partner might or will confess. Therefore, there is no choice but to confess, since confession reduces the punishment by getting a suspect closer to the bottom of his lowerarchy. He cannot get off by turning state's evidence, but he can get a lesser sentence. Since each suspect figures his partner will confess, and therefore feels he himself must confess, they are saddled with mutual confession. Any change from mutual confession by only one suspect makes his position worse, unless they both change, but this is impossible since neither trusts the other not to confess at a later date. They are locked into confessing—the moment one suspect expects the other one to confess, he himself is compelled to confess. Each has a strategy which holds the other to a fixed position (unless he is a glutton for punishment). The irony is that two prisoners unable to reason through their situation could, by not confessing, do better than those who can.

This is not a pure conflict situation between the two suspects, since it is possible for both to win at the same time and both to lose at the same time, or for one to lose more than the other wins, and vice versa.

The Prisoner's Dilemma can be related to a number of situations, some global, some common to ordinary life. For example, the United States-Soviet Union arms race is sometimes described in these terms. One country adopts a new weapons system, a MIRV system for example, with the idea of gaining a competitive advantage—"In national defense we shall be second to none!" But each time one country adopts a new weapons system, the other is compelled to equalize. (This corresponds to the two prisoners confessing.) Once this happens, both sides are worse off than they were before— each side has squandered money without gaining an ultimate advantage. However, the fear that one side might adopt a system compels the other side to adopt its own, which guarantees that the former will react likewise.

However, due to the time lag involved in developing a system, any side that feels itself behind tries not only to equalize but also to make its new system bigger and better. The other side, through its intelligence network, perceives its opponent at work on an even more advanced system than its own, confirming its original fears and spurring it to even greater effort. Built into the arms race is the question of obsolescence—the moment one country creates a new system, the other country's system is obsolete; they then have to develop a new system. In catching up with the technology of the moment there are advantages which outweigh the other country's system. The other country now must catch up with technology and create a new system. Furthermore, there is no parity because each side views parity in the most conservative possible way. What appears to one side as an attempt to gain parity appears to the other side as an attempt to gain an advantage. Since each side may see parity achieved by different technological means, the other side views these different means as an attempt to get an advantage.

Another example of the Prisoner's Dilemma is a gas station price war. Suppose there are several gas stations in a given vicinity, none of which is owned by the same company as any other. Some may be independent and others part of different chains. Suddenly, one gas station cuts the price. To keep their business, the other stations follow suit. The chain stations there react by cutting their price below cost. The independent operator must now cut his and operate at a loss. Ultimately, he may be driven out of business. He is trapped into the price war once it gets started, whether he starts it or not.

The big companies, of course, can come to an illegal price agreement among themselves, but they cannot do this so long as the independent operator can cut the price. The big companies must first start a price war to drive out the independent operator before they themselves can come to a secret price agreement (secret because such agreements are illegally ''in restraint of trade''). Obviously, the independent operator is foolish to start a price war (although many do anyway).

However, if the big companies start the price war, the small operator has no choice but to join battle—a classic example of cut-throat competition.

19 THE DOCTOR'S DILEMMA

The Prisoner's Dilemma sets up a situation which is always to the mutual disadvantage of the decision makers. Another example is malpractice insurance for doctors, the cost of which escalates each year. Naturally, the surgeon passes this cost on to his patients. The escalating cost of medical treatment is resented by the patient, who expects more for his money by demanding greater perfection and skill from the doctor—"At these charges he'd better be a genius!" In fact, in forcing the doctor to protect himself with the added cost of insurance, the patient ends up getting the same or even less for his money. If the cure fails and the patient can prove any sort of negligence, he takes the doctor to court. This causes the cost of malpractice insurance to increase, which causes the cost of medical treatment to increase.

If Irate Patient did not sue Dr. Emdee, a surgeon, then Dr. Emdee would not have to carry malpractice insurance and would not have to pass the cost of the insurance on to Irate Patient. If there were no malpractice suits, Dr. Emdee would not have to run more and more "unnecessary" tests, which are made to guard against malpractice suits. A malpractice suit can only be won by proof of the doctor's negligence, thus the doctor must cover himself with every conceivable test against the contingency of a malpractice suit. Who is to blame? The doctor? The patient? Neither?

Consider the "unnecessary" test. (We put the word unnecessary in quotation marks because few doctors will or can admit to it as such—to do so may be grounds for a malpractice suit.) Irate Patient has the options of suing the doctor or not suing him, the doctor has the options of giving "unnecessary" tests or not. If the patient sues and no tests were given, we may assume that the patient wins the case, which is best for him and worst for the doctor. If the doctor orders the tests and the patient does not sue, the doctor wins his own peace of mind but the patient loses the cost of the "unnecessary" tests. For the patient, who has been charged a high fee and is not cured—utter frustration. If the patient sues and the doctor ordered the tests, neither wins, but at least the patient has tried. But if Irate Patient does not sue, and Dr. Emdee does not order the "unnecessary" tests, everyone will be reasonably happy (Figure 28).

Irate Patient's lowerarchy:

4.   Doesn't sue/tested (Patient utterly frustrated)
3.   Sues/tested (Patient tries)
2.   Doesn't sue/not tested (Patient content)
1.   Sues/not tested (Patient wins)

Dr. Emdee's lowerarchy:

4.   No test/Patient sues (Doctor's disaster)
3.   Test/Patient sues (Doctor vindicated, but at cost)
2.   No test/no lawsuit (a square deal)
1.   Test/no lawsuit (money in the bank)

Consider the rationale for Dr. Emdee carrying malpractice insurance and Mr. Irate Patient suing, which is a quasi-Prisoner's Dilemma situation (Figure 29). We assume that Dr. Emdee would prefer the situation where Patient does

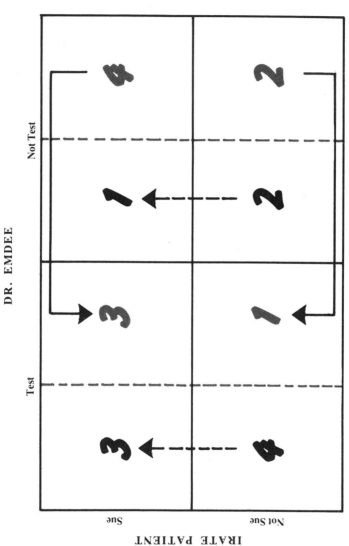

DR. EMDEE

IRATE PATIENT

FIGURE 28

NOTE: The arrows show that either Emdee or Patient always has an incentive to change strategy except when Patient sues and Emdee tests.

DR. EMDEE

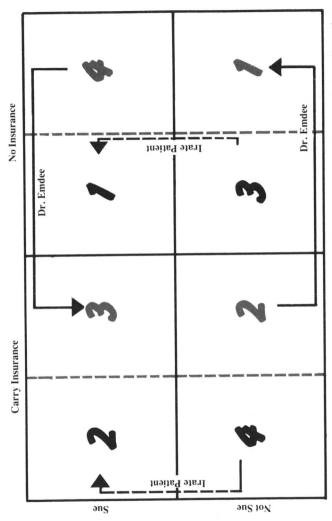

NOTE: The arrows show that at least one player always has an incentive to change strategies except when Patient sues and Emdee insures.

FIGURE 29

not sue and Emdee does not have to carry insurance. But for Patient, the only thing worse is not suing when Emdee has insurance. Patient most prefers to "sue an uninsured quack," as the outraged Irate Patient puts it. For Irate Patient, it is a moral issue. An insured doctor doesn't necessarily "learn his lesson." This is, of course, the worst outcome for Dr. Emdee. The worst for Irate Patient is not to sue when Dr. Emdee carries insurance. "Why didn't I sue him! Jesus, it wouldn't have cost him anything! And at least I could have gotten some of my money back that I've already paid for his malpractice insurance." The remaining outcome, suing when the doctor carries insurance, is for Irate Patient second only to suing when Emdee doesn't carry insurance, and for Dr. Emdee the only thing worse is being sued when he doesn't have insurance. The entire folly is summarized in Figure 29.

Irate Patient's lowerarchy:

4. Not sue/Emdee insured (missed the gravy train)
3 Not sue/Emdee uninsured (why didn't I ruin him?)
2. Sue/Emdee insured (caught the gravy train)
1. Sue/Emdee uninsured (now he works for me!)

Dr. Emdee's lowerarchy:

4. No insurance/Patient sues (which way to the Golden Gate Bridge?)
3. Insured/Patient sues (I'll have to raise my fees)
2. Insured/Patient doesn't sue (he's a nice guy, but a little stupid)
1. No insurance/no lawsuit (heaven!)

The only stable outcome is for Irate Patient to sue and for Dr. Emdee to carry malpractice insurance. Suppose Patient decides to sue and Emdee decides to carry malpractice insurance. If either changes his mind, he will increase his punishment, thus they hold each other in fixed positions. No other outcome has this property, as the reader can verify for himself by studying Figure 29. Merely pick another box besides the upper-left hand corner and see if by changing strategy at least one player can do better. For example, if Emdee carries no insurance and Patient does not sue, Patient can do better by changing his mind and suing. Thus there is only one stable outcome to the game, and the cost of malpractice insurance continues to escalate. One way to beat the Doctor's Dilemma is to eliminate malpractice suits, but this would presumably require abandoning private enterprise medicine as it exists in the United States.

20 PRISONERS OF LOVE

Sometimes a quasi-Prisoner's Dilemma situation is created unintentionally in the minds of the participants. A seemingly friendly, helpful remark can lead to unintended consequences.

Carl, a high school senior, borrows his father's car one evening to take out his girlfriend. The following morning driving to his office, the father stops for gas, and empties his ashtray. In doing so he finds a marijuana roach. He thinks about it all the way to the office. He knows that his son occasionally smokes grass and has talked it over with him. The father does not approve, but he knows he can't do anything about it. He also knows that Carl is very careful, and would never make the foolish mistake of leaving a roach lying around where it might be found. The father is confident of his son's good sense and, rather than bring about a major family crisis, he has forbidden his son to have any grass in the house. This prescription is absolute. His son must do as his conscience dictates, but the father hopes Carl will have enough integrity to resist the temptation of smoking grass on social occasions when it is passed around. He figures, therefore, that the roach must have been left by Carl's girlfriend, Carol.

The father is worried. Unfortunately, he cannot get to

148

his son until after three o'clock, when school is out. He feels he must talk to somebody about it, and phones his wife, who is out, so he anxiously phones Carol's father at his office.

He says to Carol's father, "I think Carl took Carol out last night."

Carol's father answers, "Yes, he called for her at eight. Is there anything wrong?"

"Yes. It may or may not be serious; however, I feel pretty nervous about the whole thing. This morning, on the way to the office, I cleaned out the ashtray in my car—fortunately the attendant didn't do it—and I found a marijuana cigarette butt. I'm positive it was that. Can you imagine if the attendant had found it! I never touch grass; I mean I'm absolutely against the whole thing. They must have smoked it last night. Now, I've forbidden Carl to have any marijuana in the house, and I've tried to tell him not to use it. If my wife or I had been caught with it in the car, if one of us had been stopped by the police for some minor infringement, and they had searched the car—you never know these days—and the police had found it, we would have been arrested.

"I'm afraid that, regretfully, there must be some problem here with Carol, because Carl is perfectly well aware that he must never leave any indiscreet evidence around—I've drilled it into him. I disapprove of his smoking marijuana. I disapprove of his going to parties where his friends smoke marijuana. However, I happen to know that it goes on all the time, and I'm sure you do, too. But what worries me is this: Evidently Carol must have left it in the car. She was with Carl when he was driving, and instead of throwing it out into the street, she put it into the ashtray. Leaving it in the car endangered my wife, my son Carl, and myself. Therefore, I would be very grateful if you would speak to Carol. I assume that you also disapprove."

"Yes, of course I disapprove. I'm disturbed to hear about this. I promise you I'll discuss this matter most seriously with Carol. Thank you very much for the call."

That evening Carol's father speaks to his daughter:

"You were out with Carl last night, and Carl's father phoned me this morning. Evidently he found a marijuana cigarette butt in his car. He blames you, because he says that Carl would never leave a roach in the car."

"I didn't leave a roach in the car."

"You don't smoke marijuana?"

"Well, people smoke it at parties and stuff. Carl sometimes smokes it. Lots of kids smoke grass, but I just watch it and don't have anything to do with it. I know all about marijuana and I know you don't like anybody to have anything to do with it."

Carol leaves her father with the impression that Carl's father is wrong about his son. He suspects that not only has Carl's father got the wrong end of the butt, so to speak, he is also trying to blame Carol for Carl's actions. Carol's father thinks to himself: "They [Carl and his coconspirator father] are trying to blame Carol for Carl's misbehavior, his use of drugs. I can't allow my daughter to be used this way; I've got to protect her."

He talks to Carol again: "Carol, I have no doubt you are right that Carl is on dope. I don't want you to see him anymore."

Carol absolutely refuses. He pleads with her for her own protection to stop seeing Carl. "I have had nothing against the boy, I always thought he was a fine boy, but he may be on the road to becoming a dope addict. Even if he can control his use of drugs, and doesn't take the next easy step to heroin, he is acting recklessly. I definitely don't want you to be around a person who is acting recklessly. You could get yourself into a lot of trouble, and it's just so foolish to ruin your life this way. I forbid you to see Carl anymore."

"You can't do that; I like Carl, a lot!"

Carol's father realizes there is nothing he can do to keep Carol from seeing Carl. He is afraid that Carl will be caught using God-knows-what drugs, and Carol will be caught with him.

To protect his daughter, he goes to the local police station to report that his daughter is going out with a boy who is using drugs, and that his daughter is absolutely innocent. He tells the desk sergeant the facts as he knows them, without revealing Carl's name, at first. The police, of course, figure that if the father is so concerned, the true story must be much more serious than a mere roach in the ashtray.

The sergeant says, "We're very sorry, we would like to protect your daughter, too. And we believe that the action you have taken is right, in the best interests of your daughter, and we wish more people would do this. However, we can do nothing unless we can get hold of the boy and talk to him. You know, it's in his best interest, too. One of these days, your daughter and this hopped-up kid are going to be caught. When that happens, and you better believe it will, your daughter will be brought to this or another police station, fingerprinted, photographed, booked as a common criminal, and put into a jail cell with other criminals. You'll get a call to come down to the station to talk to her. By then it will be too late. Do the smart thing now, and tell us the name of the boy. We'll go see him, talk to him, talk to his parents, and see if we can put a stop to it without anybody getting into trouble."

Carol's father tells them who Carl is.

The police get a search warrant for Carl's house, with the ultimate aim of getting information as to who is Mr. Big at the high school. They go to Carl's home, show his father the search warrant, and search Carl's room, where they discover a dried-up old joint in one of his bureau drawers. Carl is arrested and taken to the station, his father accompanying him. The police search Carl, but find no more drugs. They demand to know from whom he obtained the marijuana. Naturally he keeps his mouth shut. Carl's father says to the cops, "Look, let me talk to him alone; I'll find out what his problem is—why he won't tell you where he got it."

The son talks to his father, "Dad, what happened?"

"Carl, they've got you dead to rights, in possession. I've got to get a lawyer. Don't say a word until I get a lawyer, but I want to know what's going on. Did you buy the joint at school from somebody or is some outside pusher supplying it? Let me know who that was so I can go and tell the lawyer."

"Dad, I can't tell you; I can't tell anybody. I won't tell anybody."

"Son, you're in a dangerous situation here."

"I don't care. I just don't know what's going on. Why should they come to me?"

Then the father explains that he phoned Carol's father and it is obvious that Carol's father has turned him in.

Carl responds, "That's the trouble, I got it from Carol, and that idiot turns me in. She's got half a lid of grass; she always has it. I don't know where she gets it from, but I'm not going to turn in my girlfriend to the fuzz."

"Carl, don't say one word, but be very polite. I'll go get a lawyer."

The father realizes he's been had by Carol's father. He thinks, "My kid just doesn't know what in hell he's doing. He's going to take the rap for that dumb chick and her idiot father, who's just a self-righteous fool. I do a friendly thing, phone him, warn him that his daughter is smoking. It's obvious she's a pothead. More than that, he's too stupid to realize it, and she doesn't have the common decency to admit it. She'll let my son rot in jail. That idiot has turned my son in, when his own daughter is to blame. Maybe he even realizes that she's smoking. My son is going to take the fall, turned in by the man I was helping!"

As the only way to help his son, he is forced to tell the police about Carol. He explains that he has spoken to his son, and found out that he won't talk because he got the grass from his girlfriend. "I think you already know her name, so I don't have to tell you it."

The police get a further warrant to search Carol's

home, and, indeed, find a small supply of marijuana in her room. They take Carol to jail, accompanied by her father. Carl, when asked by the police if Carol supplied him the grass, specifically denies it. Carol refuses to say where she got the marijuana but admits giving Carl one joint as a birthday present.

The two youngsters, after a threat from the police of severe action if they ever use marijuana again, are released into their parents' custody. The two parents console each other by blaming the whole incident, not on their own stupidity, but on the lax morals of the younger generation.

The initial mistake which caused the whole incident was that of Carl's father, who unwittingly set up a quasi-Prisoner's Dilemma (Figure 30).

Carol's father's lowerarchy:

4. Not turn Carl in/Carl's father turns Carol in (he's scapegoated my daughter!)
3. Turn Carl in/Carl's father turns Carol in (pass the aspirin)
2. Not turn Carl in/Carl's father doesn't turn Carol in (a shaky situation)
1. Turn Carl in/Carl's father doesn't turn Carol in (a real gentleman)

Carl's father's lowerarchy:

4. Not turn Carol in/Carol's father turns Carl in (he's scapegoated my son!)
3. Turn Carol in/Carol's father turns Carl in (I owed it to my son)
2. Turn Carol in/Carol's father doesn't turn Carl in (my son is innocent)
1. Not turn Carol in/Carol's father doesn't turn Carl in (the best of all possible worlds)

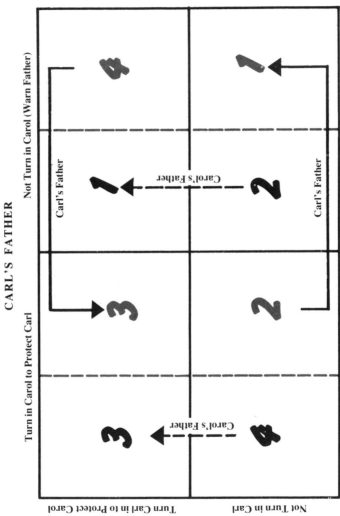

CARL'S FATHER

Turn in Carol to Protect Carl   Not Turn in Carol (Warn Father)

CAROL'S FATHER

Turn Carl in to Protect Carol

Not Turn in Carl

NOTE: The arrows show that at least one father always has an incentive to change strategies except when both kids are turned in.

FIGURE 30

Carl's father's attempt to quietly signal to Carol's father was misinterpreted; Carol's father saw this signal as a threat to his daughter's security. He invented in his own mind a Prisoner's Dilemma, in which he saw his daughter's security imperiled. To forestall this threat, he rushed to the police station, where his fears were played upon by the desk sergeant. Carol's father felt he had no choice, given his duty as a father, but to turn in Carl. Carl's father, seeing his son's security imperiled because he was in custody, realized that his only choice in protecting his son was to turn in Carol. So both got nailed.

The main ingredient to avoid being trapped in the Prisoner's Dilemma or quasi-Prisoner's Dilemma is trust. Carl's father, by mentioning marijuana, created fear and suspicion in his phone call to Carol's father. He should never have phoned him in the first place. He should have worked through Carl. Each father naturally wants to protect his child from trouble. Any time a criminal issue threatens the security of two minors from separate families there is the potential of a latent Prisoner's Dilemma, which only needs a trigger to become real.

Get your signals straight or you will fulfill your own prophecy.

Had the two fathers drawn up their lowerarchies and conflict maps before taking action, both would have realized the need for joint action to protect their interests.

21 A PERSONAL PORTFOLIO

The process that we have considered so far consists of analysis of raw emotional data applied to a game theory structure. But what brings about any crisis in the first place? Perhaps some underlying factor exists in the way you approach the world itself, one which you have no real awareness of? A typical example would be a person who talks too much; the hidden cause may be extremely petty but the consequences humiliating and time-consuming. If Colonel Blotto had kept his mouth shut and not grumbled about Liz to the committee chairman, thus raising doubts in the chairman's mind, Blotto wouldn't have provoked a crisis in the first place. He put himself on the spot unnecessarily.

One way to deal with possible underlying causes of crises is to file your conflict maps in a personal portfolio. After a while, look over the entire portfolio. See how often you are your own opponent, as in "Flashpoint!"; if so, this is not necessarily bad. Contrary to often expressed opinions, playing against yourself may be a useful way to make decisions—forcing you to look before you leap. It is as if you consciously split yourself into two persons for purposes of an internal (and effective) decision-making process, overpowering the opposition in your own mind.

Note how many of your conflict maps involve circular lowerarchies, where you had to flip a coin or use some other randomizing or scheduling device to get around the circularity. Circular conflict maps usually indicate that you are trying to achieve two different results or values which the situation may not allow you to reconcile. Since you want both, what you should do is not obvious. Randomization, scheduling, and splitting the decision are ways to reconcile the two. In "To Be or Not to Be a Pawn . . ." Fred wants to maximize his long- and short-term security at the same time; this is what gives him a circular lowerarchy. In "V.D.—Light and Easy," May tries to maintain a relationship and her freedom at the same time; this is what gives her a circular lowerarchy. In "Showdown with Colonel Blotto" Liz wants to hold onto two different projects at the same time—another circular lowerarchy.

The use of the pure strategy minimax is also a way of making a decision, but it necessitates abandoning one of the values—for instance, in "What Might Have Been," Irving abandons the chance of a spontaneous recovery. If your lowerarchies focus on only one goal, the conflict map will usually have a pure (clear-cut) good strategy. The value of the conflict map in these situations is essentially that of sorting and selecting. In "The Trouble with Harry," the conflict map allows for an intelligent weighing process. In "The Hidden Adversary—Liz versus the Jury," Liz ultimately decides that given the likely verdict of the jury, her own privacy is of greater value than fighting for justice.

Notice that the conflict maps "forget" the underlying emotions that went into forming the lowerarchy. This is a valuable asset: You now have a record of the conflict unencumbered by the anxieties that went into constructing the lowerarchy, thus making easier the search for an underlying thread connecting the conflict maps.

On analysis of your personal portfolio you may well be able to predict your next crisis and avoid it. One often hears the advice ''Don't look back,'' but sometimes looking back is a way of looking forward, especially if one examines the past pattern of events. ''Those who forget the past are doomed to repeat it,'' is another piece of everyday wisdom. The problem, however, is that this phrase gives no clue as to how much of the past one is supposed to remember. With the conflict maps, one only remembers the essential details. The trauma that could interfere with intelligent decision making is, if not forgotten, at worst only a hazy memory.

The ultimate strategist uses strategy as little as possible, because his or her life is going in the desired direction, and thus he or she is not constantly thrown into a state of crisis. Of course, one can't anticipate everything. Crises, not of one's own making, can still sneak up on even the most artful strategist, but these are the only ones a master strategist needs to confront.

Good luck!

BIBLIOGRAPHY

JOHN VON NEUMANN AND OSKAR MORGENSTERN. *The Theory of Games and Economic Behavior*. New York: John Wiley and Sons, Inc., Science Editions (paperback), 1964. Originally published, Princeton University, 1944. This is the classic work in the field. Although highly mathematical, the book was written with the aim of appealing to social scientists with little or no mathematical background. One year of college algebra (or even one semester) would be sufficient background. The effort is well worth it, even now, thirty years after publication. The authors lead the reader by the hand through the intricacies of the mathematics, and devote much of their writing to the motivation underlying the theory. The first forty-five pages should be read by anyone interested in game theory, psychology, or social science, regardless of mathematical background.

R. DUNCAN LUCE AND HOWARD RAIFFA. *Games and Decisions*. New York: John Wiley and Sons, Inc., 1957. This is the standard reference work on the subject. This book is largely nonmathematical. The section on the Prisoner's Dilemma is perhaps the most often cited section of the book.

J. WILLIAMS. *The Compleat Strategyst*. New York: McGraw-Hill, 1953. An amusing book focusing on imaginary applications of game theory. Computational methods for larger conflict maps than those we have considered are presented. Indeed, the book is largely computational and does not focus on the essential problem in application—filling in the conflict maps with low-erarchies. However, this is a very useful book for the non-mathematician.

J. C. C. MCKINSEY. *Introduction to the Theory of Games*. New York: McGraw-Hill, 1952. This is an excellent mathematics textbook on game theory.

THOMAS SCHELLING. *The Strategy of Conflict*. New York: Oxford University Press, 1963, Galaxy Book (paperback). This was a very influential nonmathematical book among political scientists in the mid-sixties. Since the focus is largely on international politics, the book contributed to the popularly held link between game theory and war. However, the book is interesting and worth reading. The thrust of the book is on "tacit" bargaining in games such as the Prisoner's Dilemma.

ANATOL RAPOPORT. *Two-Person Game Theory*. Ann Arbor: University of Michigan Press, 1966. This is an interesting book on the subject.

W. ROSS ASHBY. *An Introduction to Cybernetics*. London: University Paperbacks, 1964. This is a good introduction to cybernetics. What is cybernetics? Machines fall apart; animals rot. They also get confused. To counteract these tendencies, all animals and some machines employ principles of self-regulation. The aim of the science of cybernetics is to master these principles. Game theory, which is concerned with strategies that anticipate destructive opposition, is clearly part of cybernetics.

JAN GECSEI. "Optimal strategy generation by means of neuron networks." *Proceedings of the International Congress on Cybernetics*. Namur, Belgium, 1964. In this paper, the computational technique described in "How's the World Treating You?" is linked to the theory of neural networks.

DARREL HUFF. *How to Lie with Statistics*. New York: W. W. Norton & Company, Inc., 1954. This is a very amusing book about statistics. The topics brought up in our paragraph on averages are discussed at length.